Patrick Moore's
Practical Astronomy Series

Springer
London
Berlin
Heidelberg
New York
Barcelona
Hong Kong
Milan
Paris
Singapore
Tokyo

Other titles in this series

How to Observe the Sun Safely

Lee Macdonald

With 46 Figures

Springer

British Library Cataloguing in Publication Data
Macdonald, Lee
    How to observe the sun safely. – (Patrick Moore's practical
    astronomy series)
    1. Sun – Observers' manuals
    I. Title
    523.7
ISBN 1852335270

Library of Congress Cataloging-in-Publication Data
Macdonald, Lee, 1973-
    How to observe the sun safely / Lee Macdonald.
        p. cm. – (Patrick Moore's practical astronomy series,
        ISSN 1617-7185)
    Includes index.
    ISBN 1-85233-527-0 (alk. paper)
    1. Sun–Observers' manuals. 2. Sun. I. Title. II. Series.
    QB521 .M23 2002
    523.7–dc21   2002070456

Patrick Moore's Practical Astronomy Series ISSN 1617-7185
ISBN 1-85233-527-0 Springer-Verlag London Berlin Heidelberg
a member of BertelsmannSpringer Science+Business Media GmbH
http://www.springer.co.uk

© Springer-Verlag London Limited 2003
Printed in Great Britain

Typeset by EXPO Holdings, Malaysia
Printed and bound at the Cromwell Press, Trowbridge, Wiltshire
58/3830-543210 Printed on acid-free paper SPIN 10842242

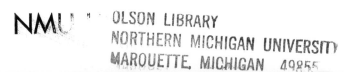

Contents

Introduction

The Sun is the brightest and most accessible object in the sky, and it has much to offer the amateur astronomer with modest equipment. On most days it shows sunspots and other features that display a wealth of fine detail and change their appearance strikingly from day to day. But observing the Sun can be dangerous. **NEVER look at the Sun through a telescope or other optical aid, even for a brief instant**. The Sun's intense radiation, amplified and focused by a telescope, will almost certainly cause eye injury and could well lead to complete blindness. Do not attempt **any** solar observing until you have read and understood the safety precautions and observing advice set out in Chapter 2 of this book – even if you think you have the correct equipment. Be especially wary about using filters to observe the Sun. If you have a filter that makes the Sun look dark, it is not necessarily safe, as it is largely the Sun's invisible radiation that is harmful to the eye.

But provided you use the correct techniques, such as projecting the solar image onto a screen or using a specially designed, quality solar filter that fits over the telescope aperture, it is quite easy to observe the Sun safely. One of the joys of solar observing is that useful observations are possible even with very small telescopes – such as the small refractors, Schmidt–Cassegrains and Maksutov telescopes that are readily available off the shelf. In fact, due in part to the fact that the Sun has more than enough light, a small telescope will actually give better results than a large one! Observing the Sun is also not affected by light pollution – a major advantage for the many amateur astronomers whose view of the night sky is obscured by the glow of streetlights and security lighting. The Sun can be observed from the town just as successfully as from the country.

As I shall explain in Chapter 1, solar activity affects the Earth in a number of important ways. For this reason, our nearest star is studied intensively by professional astronomers and is being monitored round the clock. The Sun is observed using space-based observatories as well as from the ground, and both the level of research and the equipment required to carry it out are far beyond the amateur's means. Therefore solar observing does not offer the potential for discoveries or major scientific contributions like some other branches of amateur astronomy, such as variable star observing or supernova hunting. But monitoring solar features and keeping careful records of them is still a useful contribution. Throughout the world many amateur astronomers systematically monitor the Sun and send their observations to solar observing organisations for analysis. Monitoring levels of sunspot activity is particularly useful, as it continues a long series of observations made with small telescopes over many years, and more observers are always welcome. In my view, solar photography is also useful, as it has considerable educational value. Professional solar images tend only to show small parts of the Sun or show our nearest star at invisible wavelengths, where its appearance is radically different from that in visible light. Amateur images, on the other hand, portray the Sun more realistically and so are more meaningful to the wider public.

Recent years have seen some major advances in equipment for the solar observer. Filters for white-light observing have improved, allowing better photography and visual observing. Hydrogen-alpha filters for viewing the chromosphere and prominences are now more affordable than ever, allowing amateurs with modest budgets to have views – and take photographs – once reserved for well-off individuals or professionals. Improvements to photographic films have made it easier to take good solar photographs. In the last few years, the rise of digital imaging has opened up yet more horizons. Digital cameras of the type obtainable from a camera store, intended for everyday photography, can be used to take excellent solar photographs, and the images can be processed and enhanced using ordinary image processing software on a home computer.

For all that, not much has been written about amateur solar observing in recent years. Books on general astronomy and even those specifically about the Sun tend to relegate practical solar observing to a single chapter, and so the techniques involved are not described in enough detail. The few books on the subject that have appeared have generally been either too technical or too specialised for the beginner. This book attempts to fill the current gap in the literature. I have written it primarily for the amateur who knows the basics of astronomy and wants to know how to go about observing the Sun. All the time I have emphasised what is possible using commercially available equipment that is easy to get hold of in most parts of the world. For this reason, I have deliberately eschewed some specialised topics, such as observing the Sun's radio emissions, which requires home-made equipment and a fair amount of technical know-how. Neither have I discussed in much detail the Sun-related topics of eclipses and the aurora. Both are major fields in astronomy by themselves and some good books on them have already been published.

Throughout the book my emphasis has been on *practical* solar observing – what *you* can do with ordinary equipment, provided you take the proper safety precautions. I have tried to avoid unnecessary theory and have not attempted detailed scientific explanations, as these are available elsewhere. Rather, this book is intended as a basic guide to give the amateur a taste for observing our ever-changing nearest star, in the hope that he or she will explore further.

Acknowledgements

I would like to thank Derek Hatch and Eric Strach for allowing me to use some of their fine solar images in the text, and Harold Hill for his masterly drawings of solar prominences. Thanks are also due to David Boyd for supplying photographs of his solar observing equipment and to Geoff Elston, Director of the Solar Section of the British Astronomical Association, for putting me in touch with some of the contributors. Last, but by no means least, I am grateful to John Watson, Astronomy Editor at Springer-Verlag, for giving me the opportunity to write about observing the Sun and for all his help in seeing the work through the press.

Lee T. Macdonald
Newbury, Berkshire
June 2002

Chapter 1

Introducing the Sun

The Sun is important to astronomers for two reasons. The first is that it is Earth's only natural "power station", producing the light and heat essential to life on our planet. Without the Sun, the Earth would be more or less a frozen ball of rock, with no atmosphere, no weather, no life and no people. The Sun can also be harmful to us, because it emits huge quantities of radiation that would be fatal to humans and all living matter, were our planet not protected from it by a thick atmosphere and powerful magnetic field. But intense bursts of solar activity can still harm communications and electrical power systems, so we need to keep a constant watch on the Sun so that we are forewarned of its next powerful outburst. We also need to understand it so that we can predict future activity and its likely consequences.

The second reason for studying the Sun is that the Sun is a star, much like the three thousand or so other stars that we can see in the night sky. All the stars in the night sky, however, are exceedingly remote from us. Even the closest known star, Proxima in the southern constellation of Centaurus, is some 60 million million kilometres (40 million million miles) from Earth – so far away that its light takes over four years to reach us. Most stars – even the majority of those visible to the naked eye – are much further away still. Even the world's most powerful telescopes show the stars as mere points of light and give us just basic information as to what they are and how they work. But at only 150 million kilometres (93 million miles) from Earth, the Sun is easily close enough for us to have a detailed view and learn much about it. Armed with a detailed knowledge of the Sun, astronomers can learn more about other stars.

The Sun's Place in Space

In terms of its size and luminosity, the Sun is an average star. Some stars are much larger and brighter than the Sun and some of the largest, the "supergiants", would extend out to the orbit of Mars or even beyond if placed in the Solar System. At the other end of the scale, the "dwarf" stars emit a relatively feeble light and are typically no larger than the Earth. The varying brightnesses of the stars that we see in the night sky give no indication of the stars' sizes and actual brightnesses, because the apparent brightness of a star is also determined by its distance from us. Some stars in the night sky appear bright only because they are relatively close to us. A good example is Sirius, the brightest star in the sky. This is, in fact, quite an average star, but it is quite close by at "only" 9 light-years. (A light-year is the distance travelled by light in one year – approximately 9.46×10^{12} km or 5.9×10^{12} miles.) However, not far away in the winter sky is Rigel at the foot of the constellation Orion. It appears only slightly fainter than Sirius, but it is 800 light-years away, making it 60,000 times as luminous as the Sun! You can get a good analogy of the difference between a star's true and apparent brightness by standing in the same room as a 40-watt light bulb and looking out at a distant floodlight or security light. The light bulb a few feet away from you appears brighter, even though the security light is a much more powerful device. As stars go, the Sun is one of the "light bulbs" – not one of the dimmest stars in our galaxy, but certainly not one of the brightest either.

In ancient times – and, indeed, until less than 500 years ago when Copernicus published his "heliocentric" theory of the universe – it was generally believed that the Earth lay at the centre of the universe and all the other heavenly bodies – including the Sun – circled round it. Astronomical research in the five centuries since then has revealed that nothing could be further from the truth. Our Earth and even the Sun are very junior members of an exceedingly complex universe. The Earth is but one of nine planets (plus countless smaller bodies) orbiting the Sun, a very average star.

The Sun is just one of 100 billion (i.e. 100 thousand million) stars in a huge, rotating disc which we call the Milky Way galaxy. The Sun is by no means near the centre of the galaxy; in fact, we are about 28,000 light-years out, in one of the spiral arms. The Milky Way, in turn, is just one of millions of galaxies scattered throughout the universe. The Sun is thus relatively unimportant in the grander scheme of things, but to us on Earth it is essential to our existence.

How the Sun Works

The Sun and other stars are nuclear powerhouses which generate their own heat energy by means of nuclear fusion reactions at their cores. The fusion process releases a huge amount of energy, and it is this energy that powers the Sun and stars. Every second the Sun converts 4 million tonnes of hydrogen into energy, which is eventually released into space. The Sun is thus continually getting lighter, but the mass loss is so small in relation to the Sun's total mass that it will have no significant effect for billions of years to come. For fusion to happen, the temperature at the core has to be very high. Current estimates say that the Sun's core has a temperature of around 16 million degrees kelvin ($^\circ$K).

The core in which the fusion reactions occur extends outwards to about a quarter of the Sun's radius. The energy produced by the core is radiated outwards to the upper regions of the Sun's interior through a very stable region of gas extending to more than two-thirds of the way to the Sun's surface. In this *radiative zone* energy from the core is constantly absorbed, re-radiated and deflected by the hot surrounding gas, so that it takes some 170,000 years for energy to travel from the core to the upper regions of the Sun. Thus the energy that we see as light on the Sun's surface started its journey from the core during the last Ice Age!

At about 500,000 km from the Sun's centre the gas becomes too cool for the radiation to flow outwards via the above process. As the radiation is blocked, pressure builds up and this causes the radiation to flow to the surface via convection currents, much as heat from a radiator or fire rises into the cooler surrounding air. Heat takes about 10 days to pass through this *convective zone*, at the end of which it finally arrives at the surface, known to astronomers as the *photosphere*, which we see as the luminous disc of the Sun. Although the photosphere appears brilliant to the eye, it is actually relatively cool, at "only" 5,800°K.

Actually, the word "surface" is a misnomer here, as it is not solid but is as gaseous as the rest of the Sun. However, the photosphere *is* a surface in the sense that it sharply defines the outline of the Sun as seen in ordinary, visible light. It is also the layer from which most of the Sun's energy is radiated.

The Sun is surrounded by a very extensive atmosphere, but its density is extremely low, making it much fainter than the photosphere. As seen from Earth, the Sun's atmosphere is totally overwhelmed, because the brilliant light of the photosphere is scattered by the Earth's atmosphere. The solar atmosphere is therefore invisible from Earth except when the photosphere is hidden by the Moon during a total solar eclipse. The lowest layer of the atmosphere is called the *chromosphere*, which extends to only a few thousand kilometres above the photosphere. It can be seen during a total eclipse as a thin band of light around the silhouette of the Moon, appearing red or pink in colour. This colour led astronomers to give it its name, *chromos* being the Greek word for "colour". It glows pink because it is composed mostly of ionised hydrogen which emits light at a distinct series of wavelengths, the brightest being the so-called hydrogen-alpha (H-alpha) line, which glows red. Because it emits at discrete wavelengths, the chromosphere can be viewed from Earth without having to wait for an eclipse, using special instruments which block out all other wavelengths. Special filters are now available which allow even amateur astronomers to view this part of the solar atmosphere, as I shall describe in Chapter 6. This is very fortunate because, as I shall explain shortly, much dramatic activity occurs in the chromosphere.

The main part of the solar atmosphere is the *corona*. During a solar eclipse this is visible as a broad ring of brilliant white light which has no definite outer edge but gradually fades with distance from the Moon's silhouette. It shines due to scattering of light from the photosphere by electrons and dust particles. One aspect of the corona has been known for a long time: it is extremely hot, with an average temperature of around 2 million degrees kelvin. The chromosphere, too, is considerably hotter than the photosphere, with an average temperature of about 10,000°K. Between the chromosphere and the corona is a thin region known as the *transition region*, in which the temperature increases by a factor of 100. The corona is so hot that it shines in X-rays and the extreme ultraviolet. This intense radiation is blocked by the Earth's atmosphere – thankfully, because it would quickly kill all living matter on Earth – and so it requires space-based instruments to be observed. Why the corona is so hot is one of the great unanswered questions about the Sun and the subject of much research by professional solar astronomers. They believe that the cause of the heating is connected with Sun's powerful, ever-changing magnetic field.

The corona has no definite outer edge, and instead thins out gradually with distance from the Sun. In fact, it continuously emits a stream of protons and electrons known as the *solar wind*. The particles fly outwards from the Sun at an average speed of 400 kilometres per second and are constantly replaced by new matter. The solar wind pervades the entire Solar System – indeed, it extends outwards well beyond the orbits of the nine major planets, and no one knows precisely where its influence ends and gives way to the much more rarefied gas between the stars.

Solar Activity

Aristotle and most other western philosophers of ancient times believed the Sun to be a ball of pure white fire. However, as early as two centuries before Christ, astronomers in China began reporting occasional dark patches on the face of the Sun when it was setting and so the Earth's atmosphere made it dim enough to look at. These were the first reports of *sunspots* (Figure 1.1), whose presence was confirmed using the newly invented telescope by Galileo Galilei and other astronomers in the early seventeenth century.

A sunspot appears dark because it is about 2,000°K cooler than the surrounding photosphere. As the temperature of the photosphere is around 5,800°K, a sunspot is still extremely hot by earthly standards and only appears dark compared with the brilliant photosphere. If it were possible to isolate a sunspot from its surroundings it would still shine very brightly. Sunspots come in all shapes and sizes – from single, isolated spots to complex groups containing anything from a few spots to over a hundred. Sunspots are forming and dying out all the time, and the Sun's appearance is never quite the same from one day to the next. The size

Figure 1.1. The Sun, showing sunspots.

of sunspots gives some appreciation of the vastness of the Sun. A typical small to medium-sized sunspot is about the same size as the Earth, while some of the largest spots have lengths ten times the Earth's diameter.

The number of sunspots present is not constant but varies in a cycle from a maximum down to a minimum and back to maximum again over a period of 11 years. Astronomers call this the *sunspot cycle* and the maximum and minimum are known as "sunspot maximum" and "sunspot minimum". At sunspot maximum the Sun is heavily spotted on most days, whereas sunspots may be entirely absent for days at a stretch at minimum. Strictly speaking, the term "sunspot cycle" refers only to sunspots. Other, less-visible forms of solar activity also wax and wane over the same period and when referring to these or solar activity as a whole astronomers use the term *solar cycle*. Why solar activity varies in this manner is another great mystery of solar science. What is known is that while its period has been around 11 years long since telescopic observations of the Sun began, each cycle has different characteristics. Some cycles are much more active than others, with more sunspots visible at maximum. Also, the graph of solar activity is different for each cycle. Sometimes the cycle takes longer to reach maximum, and activity can remain the same or even decline for several months before resuming its upward trend. One common feature of all sunspot cycles is that activity takes longer to drop to minimum than it does to reach maximum.

At the beginning of the twentieth century astronomers discovered that sunspots are strongly magnetic. Sunspots tend to form in pairs and even large sunspot groups often consist of many small spots and a pair of larger spots. Each sunspot in a pair has an opposite magnetic polarity – one spot negative, the other positive. However, the polarities are reversed in the opposite hemisphere. For example, if sunspot pairs in the northern hemisphere have a preceding spot of positive polarity and a negative following spot, then in the southern hemisphere it will be the other way round: preceding negative, following positive. These polarities switch at the end of a sunspot cycle, at minimum activity. In the next sunspot cycle, therefore, preceding spots will be negative in the northern hemisphere and positive in the southern hemisphere. The polarities do not return to their original state until the completion of the second cycle and so the Sun can be said to have a magnetic cycle 22 years long – twice as long as the sunspot cycle.

The magnetic nature of sunspots is an indicator of a fundamental property of the Sun. Our star has a huge and powerful magnetic field, which astronomers believe is powered by complex motions within the solar interior. Sunspots are believed to be caused by the magnetic field being distorted by the Sun's rotation. Near the surface the solar rotation speed is not the same all over the Sun. At the equator the solar surface rotates once every 25 days, but near the poles the rotation period increases to 30 days. This *differential rotation* causes the magnetic field lines to become tangled and gives rise to localised "knots" of very intense magnetism. A strong magnetic field stops the passage of energy and when such a field pierces the Sun's surface it prevents light and heat from welling up from the solar interior, giving rise to a darker, cooler patch on the surface – i.e. a sunspot.

A good analogy for a sunspot pair is a simple bar magnet, each end of which is of opposite polarity. If you perform the well-known school experiment of placing the magnet under a sheet of paper which has been evenly spread with iron filings, the loops of the magnetic field lines become apparent. Sunspots also have magnetic field loops around them. Although they are invisible in ordinary light, they show up in other wavelengths, especially in X-rays and the extreme ultraviolet. Images taken in X-rays by spacecraft such as the Japanese Yohkoh satellite, or in the extreme ultraviolet by the Solar and Heliospheric Observatory (SOHO) spacecraft show extremely hot gas flowing along the magnetic field lines above sunspots. The magnetic fields around sunspots are huge and, in fact, sunspots are just the tip of the iceberg in a large field of magnetic activity. For this reason, professional astronomers refer to sunspot groups as *active regions*.

In the chromosphere, interaction between magnetic fields causes a particularly energetic form of solar activity: *solar flares*. A powerful flare can release as much energy as billions of nuclear explosions. All but the strongest flares are invisible in normal light but they are very prominent at certain limited wavelengths of the visible spectrum and so can be detected from Earth using special instruments. However, most of the energy in flares is emitted in X-ray and radio wavelengths, where they can sometimes outshine the entire Sun. Astronomers believe that flares are triggered when complex magnetic fields in active regions become wound up in tight spirals and connect with lines of opposite polarity, causing a sudden release of energy. Flares most commonly occur above large, complex sunspot groups where magnetic fields of opposing polarities are often found tangled together.

During a total solar eclipse it is usually possible to see a number of pink, flame-like structures around the silhouette of the Moon. This is another form of chromospheric activity, called *prominences* (Figure 1.2) – large clouds of hydrogen shining at the same H-alpha wavelength as the chromosphere and visible from Earth outside eclipses using special filters and other instruments. Some prominences show little change in their appearance over weeks or months, while others can erupt from the Sun and disappear very quickly. Both types often show beautiful, intricate structure.

Solar activity is also revealed in large-scale changes in the corona. From time to time apparent gaps known as *coronal holes* appear in the corona. Coronal holes are areas in which the magnetic field stretches out indefinitely into space and does not loop back into the Sun. They are a means by which the Sun emits material out into the Solar System and so have some effect on Earth (see below). Coronal holes and other features of this part of the solar atmosphere are only

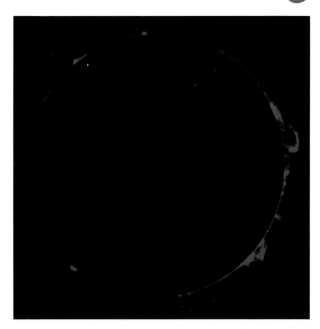

Figure 1.2.
Prominences around the whole Sun, photographed by Derek Hatch. This picture was not taken during a solar eclipse but under normal sunlit conditions using a coronagraph, an instrument employing an occulting disc to block out the Sun's disc.

visible on images taken through X-ray and extreme ultraviolet telescopes above the Earth's atmosphere. The familiar photosphere is too cool to emit X-rays and so appears as a dark sphere on such images. On the other hand the corona, at 2 million degrees kelvin, glows brightly at these wavelengths and any active regions present show up brilliantly against the dark photosphere. Coronal holes appear as dark regions with little or no activity.

Solar activity also occurs on an even larger scale in the form of huge "bubbles" of coronal gas which spew out of the Sun and become shock waves in the solar wind. These *coronal mass ejections* (abbreviated to CMEs) are usually connected with erupting prominences and are sometimes associated with flares. They broadly follow the solar cycle both in their frequency and their intensity. CMEs occur over periods of several hours and the clouds of ejected matter can occasionally become larger than the Sun's disc before dispersing. Since 1996 the large-scale structure of the corona has been more or less continuously monitored by the SOHO satellite. Time-lapse movies assembled from SOHO images of the corona dramatically show the expanding shock waves of CMEs. You can view these SOHO movies at various sites on the Internet.

The Sun's Influence on the Earth

Astronomers have long wondered whether the Sun has an effect on the Earth's weather and climate. However, after centuries of research and speculation the jury is still out on this question. We can certainly say that sunspot maxima and minima do not coincide with warmer and cooler spells on Earth. Nor does a large sunspot cause a period of unusually warm or cool weather. Most scientists now believe that the effects of solar activity are too weak compared with the main forces that drive the Earth's weather to influence weather systems. However, there is some evidence that solar activity may affect our climate over longer periods. During the period from about 1645 to 1715 Britain and Europe experienced a "Little Ice Age", in which summers were unusually cool and winters severe, the River Thames frequently icing over. Annual growth rings in tree trunks show that trees grew less during this period. This spell almost exactly coincides with a period of very low sunspot activity known as the "Maunder Minimum", in which even at solar maximum very few sunspots were seen.

Another reason for keeping a close watch on the Sun is that solar activity affects us in a number of very practical ways. The Sun's powerful magnetism strongly influences Earth's own magnetic field, which protects us from many of the potentially dangerous charged particles emitted by the Sun. The solar wind constantly pummels the Earth's magnetic field on Earth's sunward side. In the vicinity of the Earth the solar wind travels at approximately 400 kilometres per second and so compresses the sunward side of the magnetic field. Unable to get any further due to the opposing force exerted by the field, the solar wind particles are deflected round the field and eventually flow out into interplanetary space beyond the Earth, thus safely bypassing our planet. This "safe haven" in which the Earth and its inhabitants are shielded from the continuous stream of solar particles is known as the Earth's *magnetosphere*.

During periods of high solar activity, however, the Sun releases intense bursts of particles and the Earth's magnetosphere cannot keep them all out. Some particles flow down the magnetic field lines towards the magnetic poles and give rise to the aurorae – the well-known "Northern Lights" (or "Southern Lights" in the southern hemisphere) – which is basically a glow caused by the particles hitting the upper atmosphere. Normally the aurora is only visible towards Earth's polar regions, but if activity is particularly high a *geomagnetic storm* can occur, allowing the aurora to be seen at temperate and occasionally even tropical latitudes. A good display of the aurora is a memorable sight, with vivid red and green colours and ghostly patterns of luminous rays and arcs. A geomagnetic storm is caused when the Sun releases a large burst of charged particles into the solar wind and this burst reaches the Earth's magnetosphere. Flares, coronal holes and mass ejections can all cause these bursts of particles.

The frequency and intensity of geomagnetic storms peaks near the maximum of the solar cycle and bottoms out near minimum. The correlation is far from precise, however. Major storms and auroral displays visible down to temperate latitudes are not unknown near minimum, while at maximum there can be long periods without a large storm. Similarly, the presence of a large sunspot or a large number of sunspots is not reliable as an indicator of auroral activity. Remember that it is not the sunspots themselves but the *invisible* activity in the solar atmosphere – i.e. flares, coronal holes and mass ejections – that causes the fireworks. Some sunspots can cause more numerous and more powerful flares than others, while sometimes a really big flare can come from an innocuous-looking spot. Bursts of particles are not always directed towards Earth and so can miss us entirely. Also, conditions in our own upper atmosphere and magnetosphere can mean that even during a very large geomagnetic storm aurorae are only visible from high latitudes. The only way to be sure of eventually seeing an aurora is to keep checking the sky towards the north each evening.

Solar storms have more effects than colourful displays of the aurora and can affect our daily lives too. On Earth they can damage power and communications systems. For example, the great storm of March 1989, which caused a spectacular auroral display, took many people by surprise by introducing a huge amount of electric current into power distribution systems in the USA and Canada. Several generators burned out, causing six million people to be without electricity for several days. Magnetic storms can also cause serious interference in short-wave radio communications.

More serious are the effects of solar storms on the satellites orbiting the Earth. Satellites are vital to the modern world in a number of ways: navigation, defence, emergency services, weather forecasting, environmental monitoring and telecommunications. Communications satellites are vital in television, the Internet and the telephone, including mobile phones. If anything happens to a satellite, therefore, the effects could be serious. If a large flow of particles enters the magnetosphere during a magnetic storm, a satellite can be overloaded with charge and so burn out, or its on-board computers can reset themselves and so cause malfunctions. An example of such a failure occurred in January 1997 – surprisingly, just a few months after sunspot minimum – when the US communications satellite Telstar 401 underwent a power failure and broke down completely at the same time as particles from a CME hit the magnetosphere. Although we will never be totally sure that the CME was the cause of this particular failure, anomalies experienced by another satellite at the same time strongly suggests that the two events were related.

Solar activity also has a longer-term effect on satellites. During solar maximum, and especially during magnetic storms, the increased numbers of charged particles cause the Earth's atmosphere to expand outwards. This causes many satellites in low Earth orbit to lose their orbital speed due to air resistance and thus gradually decrease in altitude, an effect known as "atmospheric drag". If a satellite loses too much altitude it eventually slows down too much for it to remain in orbit and succumbs to the Earth's gravity, burning up in the atmosphere as it drops towards the ground. Atmospheric drag caused the Skylab space station to fall to Earth in 1979 and also finished off the Solar Maximum Mission satellite ten years later. Ironically, both spacecraft did important research on the Sun.

We can see, therefore, that the Sun is a highly complex object whose activity has important effects on the Earth. It is also a nearby star, from which astronomers can learn more about other stars. For these reasons, the Sun is studied intensely by large teams of professional astronomers, both using space-based telescopes and at observatories on Earth situated at the best observing sites, attempting to answer the many unsolved problems in solar astronomy, such as why the corona is so hot. But the Sun is also accessible with small and simple telescopes, and so amateur astronomers can learn much about the Sun by observing it on a regular basis. It is also possible for the amateur to make useful observations, both in monitoring levels of sunspot activity and in taking meaningful images that can help educate the general public about the Sun. Now that we have set the scene with this general description of the Sun, we can set about how to begin observing it – first, by assessing what equipment we need to observe the Sun safely.

Chapter 2

Equipment for Observing the Sun

I will begin this chapter by repeating the warning I gave at the start of the book. **NEVER look directly at the Sun through ANY form of optical equipment, even for an instant.** A brief glimpse of the Sun through a telescope is enough to cause permanent eye damage, or even blindness. Even looking at the Sun with the naked eye for more than a second or two is not safe.

Do not assume that it is safe to look at the Sun through a filter, no matter how dark the filter appears to be, because the danger of observing the Sun is caused less by its exceedingly bright light – though that would be enough to blind you if you looked at it for long enough – but by its *invisible* radiation, from the infrared and ultraviolet parts of the spectrum. When a telescope is pointed towards the Sun, a large amount of infrared and ultraviolet radiation – as well as visible light – is concentrated at the eyepiece. The infrared radiation manifests itself as intense heat – a fact that can be verified by placing your hand behind a telescope or even a magnifying glass pointed at the Sun. The heat is easily enough to cause paper to smoulder after a few seconds. When the eye is exposed to the Sun in this manner, the infrared radiation burns the retina (the light-sensitive part of the human eye), destroying the light receptors. A filter may block enough visible light, but it can still pass dangerous amounts of invisible infrared and ultraviolet without you knowing it. More frighteningly, the retina is not sensitive to pain, and so you may not be aware that your eye is being damaged while you are looking at the Sun. A retinal burn can also spread from a small hole in your vision to a large blind area over the ensuing weeks. Fortunately, there are nowadays a number of safe ways in which to observe the Sun. Do not attempt solar observing until you have read and carefully considered the advice below. Your eyes are too precious for it to be worth taking risks.

The first question is what sort of telescope is best suited to observing the Sun. The short answer is simply to use what you've got, as each of the main types of telescope used in amateur astronomy can be employed to some extent for solar observation. I will begin here by outlining the strengths and weaknesses of each type and so enable you to decide which type of solar observation your existing telescope is best adapted to – or allow you to make an informed choice if you are contemplating buying a telescope for solar study. I will then describe how to observe the Sun safely using each type of telescope.

Telescopes for Solar Observing

Unlike some other branches of astronomy, observing the Sun does not require a large telescope. While many targets in the night sky are difficult to observe because they emit too little light, this is not a problem with the Sun – indeed, the Sun has far too much light! Also, the Sun shows plenty of detail even in a very small telescope. Whereas details on planets or deep-sky objects often require a good-sized instrument to be seen, dramatic changes in sunspots and other solar features can be observed and recorded using just a 60 mm (2.4 inch) refractor, provided that it is equipped with appropriate eye protection.

Three basic types of telescopes are used by amateur astronomers: the refractor, which uses lenses to focus the light, the reflector, which uses mirrors and the catadioptric, a generic name which covers two different designs combining lenses and mirrors to form an image at the eyepiece.

The Refractor

The refractor is in many ways the best type of telescope for observing the Sun. All modern refractors use an object glass composed of at least two pieces of glass, in order to eliminate "chromatic aberration" – fringes of bright colour around the image caused by the glass splitting up light into its spectrum of colours. The commonest type of modern refractor lens is known as an "achromatic" lens. This type of lens, while greatly reducing the amount of false colour, does not eliminate it entirely and in recent years a number of other lens designs have been produced, involving three or four glass elements. Telescopes using these multi-element lenses are known as "apochromatic" refractors and they give excellent images almost entirely free from false colour. They have the advantage of having shorter tubes than achromatic refractors of the same aperture and this is especially useful in refractors of more than 100 mm (4 inch) aperture, because it makes them more compact and portable. Whereas achromatic refractors generally have focal lengths between 10 and 15 times their apertures – that is to say, they have focal ratios (abbreviated f/ratios) of between f/10 and f/15 – an apochromat can have an f/ratio of f/8 or even shorter. However, these telescopes are very expensive and, while they are perfectly suitable for observing the Sun, they are of no great advantage, as the very small amount of false colour produced by an ordinary achromatic refractor is not enough to significantly degrade the Sun's image. Apochromats come into their own in photography of deep-sky objects, because they give sharp images over a wide field of view using a relatively short time exposure. Do not buy a simple achromatic refractor with a focal ratio shorter than f/10 if you have a serious interest in solar observing. These short-focus refractors were designed for visual deep-sky observing and are not suitable for the rather higher magnifications employed in observing the Sun.

A small achromatic refractor, with an object glass of between 60 mm (2.4 inches) and 100 mm (4 inches) and a focal ratio of between f/10 and f/15 has a number of distinct advantages for the amateur solar observer. Firstly, such a telescope is relatively cheap to buy, the prices of 60 mm refractors starting at as little as £100. The second main advantage of a small telescope such as this is portability. Ideally, a telescope used for serious solar observing should be permanently set up in an observatory. However, for it to be worth building an observatory requires a site that has access to most of the sky and where objects low on the horizon can be observed. Most of us do not have a suitable site, let alone the money to construct a good-quality building. Even if you are lucky enough to have an observatory, the Sun might not be accessible when you want to observe it. The Sun's position in the sky varies greatly with the time of day and the time of year. For example, it may still be behind the rooftops in the south-eastern sky when you want to have a quick look at the Sun on a winter's morning before going to work, or it could be equally inaccessible in summer evenings, when the Sun is in the north-west. For reasons I will describe in Chapter 3, there are a lot of advantages in observing the Sun when it is between 10 and 30 degrees above the horizon and so it is a good idea to have a telescope that is compact and light enough to be carried to a spot where the Sun can be seen. Unlike a larger telescope, a small instrument can be carried in one piece and can be used at a moment's notice – an essential requirement if

Figure 2.1. The author's 80 mm (3.1 inch) refracting telescope on a German equatorial mount with electric slow motion controls. A projection box is attached to the eyepiece end of the telescope for safe viewing of the Sun's image.

you live in a cloudy climate, where every minute of clear sky is precious.

Another major advantage of a small refractor is that it does not collect much heat. This is of great importance if you are using the telescope to project the Sun's image onto a screen and you are not employing a filter over the telescope's front aperture. When focused inside the telescope, the Sun's rays generate a great amount of heat which can cause turbulence in the air inside the tube and so cause the solar image to flicker and blur. Even more importantly, the concentrated heat inside a medium or large-sized telescope can be enough to damage your telescope's internal components – particularly if, as with some modern telescopes, the tube contains parts made of plastic. Reflectors and catadioptric telescopes both have components close to the focus and so are very susceptible to heat damage. Even refractors, however, can collect dangerous amounts of heat, and some of the cheaper modern refractors have plastic drawtubes and other internal parts. Heat damage caused by projecting an unfiltered image could damage the warranty in some telescopes. But by following the precautions I shall describe below in the section on eyepieces for solar observing you should be able to avoid heat problems. I have used 80 mm (3.1 inch) and smaller refractors for over a decade with no problems at all (Figure 2.1). Small refractors also allow you to easily attach your own accessories for projecting the Sun's image, as I will describe below.

The Reflector

The reflector is more strictly known as the Newtonian reflector, after its inventor, Sir Isaac Newton. In many ways it is the best all-round telescope for the amateur astronomer. Of the three main types of telescope, the Newtonian is the cheapest per unit aperture. Because Newtonian telescopes are cheaper, they tend to be sold in larger apertures than refractors. A 114 mm ($4\frac{1}{2}$ inch) Newtonian is usually the smallest size commercially available and the 150 mm (6 inch) size has long been popular with the average amateur astronomer. A good 150 mm Newtonian can often outperform a smaller but more expensive refractor in picking up faint star clusters and galaxies and can give very satisfactory results on the Sun too. However, it is less convenient than the refractor for solar work, for several reasons. The larger size of a reflector makes it less portable and the large aperture is of no advantage in solar observing. While in theory a larger telescope should be able to resolve more fine detail on the Sun than a smaller instrument, the atmospheric turbulence caused by the Sun heating the ground during the day rarely permits this. In fact, the large aperture can be a danger, due to the concentrated heat inside the tube making the telescope effectively a solar furnace. The secondary mirror or "flat" in a reflector is especially susceptible to overheating, since it lies quite close to the focus. Even if the heat inside the tube is not great, it can still be enough for the secondary and its mount to generate a small current of warm air, spoiling the solar image. These tube currents are worsened by the fact that a Newtonian has an open tube. This allows the warm air to mix with the cooler air outside, causing further instability. You can overcome the heat problem by covering the front of the telescope with a cardboard mask with a small hole cut in it. For a 150 mm reflector, the hole should be about 60 mm (2.4 inches) in diameter and it should be positioned away from the centre of the tube, so that the secondary mirror or its support vanes do not block the light path. Such an "off-axis mask" greatly reduces the amount of heat in the tube but still allows plenty of detail to be seen on the Sun.

Catadioptric telescopes

Since they first appeared on the amateur telescope market in the 1970s, catadioptric telescopes have become very popular with amateur astronomers. Two designs of catadioptric telescope are readily available: the Schmidt–Cassegrain telescope (commonly abbreviated to SCT) and the Maksutov. Both designs work on the same principle: the light passes through a specially designed corrector lens at the front of the tube before the main mirror reflects it onto a much smaller convex mirror set just inside the corrector lens. The convex mirror then reflects the light back down the tube through a hole in the main mirror and the image is focused and viewed at the bottom of the tube, as in a refractor. The main physical difference between the two designs is the shape of the corrector lens.

Because they reflect the light twice along the length of the tube, catadioptric telescopes are very compact and lightweight. This makes them very portable and also easy to mount rigidly. The latter quality means that SCTs and Maksutovs are well-suited to astronomical photography and CCD imaging and they are marketed for these purposes as well as visual observing. Telescopes of this type designed for serious imaging of the night sky are available in apertures of 200 mm (8 inches) and above, but catadioptrics designed for visual observing and short-exposure photography are also available in smaller apertures. Widely advertised models of these small telescopes include the Meade ETX range of Maksutovs (available in apertures from 90 mm (3.5 inches) up to 125 mm (5 inches) – see Figure 2.2), the 90 mm Maksutov made by Questar and various Schmidt–Cassegrain and Maksutov instruments offered by Celestron.

All catadioptric telescopes can be used for solar observing, but they *must* be used with a filter. These telescopes contain some delicate internal components and the Sun's heat could easily damage them if they were used for projection. In any case, the mountings of these telescopes would make it difficult to attach projection apparatus to them. Fortunately, many front-aperture filters are available to fit specific models of SCT and Maksutov telescopes. These instruments are also espe-

Figure 2.2. A Meade ETX 90 mm (3.5 inch) Maksutov telescope equipped with a glass aperture filter. Note that the finderscope has been removed. This is an important safety precaution, because the Sun is just as dangerous to look at through the finderscope as the main telescope and careless people, particularly children, could look through it by accident.

cially suitable for solar photography, as they were designed with astronomical photography in mind and many adapters and attachments are available for them.

Mounts

Just as important as the telescope itself is the mount that supports it and moves it to track the apparent motion of the Sun. A telescope on an inadequate mount is very frustrating to use, as the image constantly dances about when the telescope is adjusted or focused. Even vibrations from wind or passing traffic can cause annoying vibrations in a telescope on too flimsy a mount. Many very cheap, small telescopes sold in department stores and high street shops are on very poor-quality mountings and these telescopes – which are optically inferior anyway – should be avoided.

Mountings for both amateur and professional tele-scopes are of two basic types – *alt-azimuth* and *equatorial*. The simplest is the alt-azimuth, in which the telescope moves round horizontally (azimuth) and up and down vertically (altitude). An equatorial mounting is effectively an alt-azimuth mounting that has been tipped over so that the azimuth axis points towards the celestial pole. Alt-azimuth mountings for small refractors are on tripods and the telescope is carried on either a small fork or a pan-and-tilt head similar to that on a camera tripod. Alt-azimuth reflectors are usually on "Dobsonian" mountings, a form of alt-azimuth mount which runs on Teflon bearings and has a low centre of gravity, giving a very stable mount with smooth motions for a reflecting telescope's relatively bulky tube. Equatorial refractors and Newtonians normally employ the "German" mount design (see Figure 2.1). One disadvantage of this design is that it requires a counterweight to balance the telescope, thus adding to the weight of the overall setup. Catadioptric telescopes are usually on "fork" mountings, an equatorial design suitable only for telescopes with short tubes but which dispenses with the counterweight, making such telescopes even more portable.

An alt-azimuth mount is the cheapest option and for a refractor or Newtonian owner it is also the most portable. General solar viewing, sunspot counting and even basic drawings and photography can be done with a telescope on a good alt-azimuth mount. As you become more experienced, however, you may wish to upgrade to an equatorial mount. The main disadvantage with an alt-azimuth mount is that in order to keep the Sun centred in the field of view you need to use two slow motion controls – altitude and azimuth. An equatorial mount allows the Sun to be tracked with only one motion. Best of all is an equatorial mount with a motor drive to keep the Sun in the field of view. Also, an equatorial mount that is properly aligned with the pole enables the orientation of the Sun's image to remain the same throughout the period in which it is observed – an essential feature for serious sunspot counting and drawing the Sun's disc to determine the positions of sunspots. With an alt-azimuth mount you need to adjust the orientation every few minutes to ensure accurate results. Electronic slow motions are a very useful accessory for a motor drive, as they allow you to scan the Sun at high magnification and make tracking adjustments.

Some modern telescopes have alt-azimuth mounts with motorised tracking, the positional adjustments being made by the instrument's internal computer. However, although these mounts keep the Sun conveniently in the field of view, they do not correct for the changing orientation of the image, and so you still need to adjust the image to keep it aligned. In any case, computer-controlled alt-azimuth mounts are generally sold with "Go To" telescopes designed for automatically finding faint objects in the night sky. While these telescopes are excellent for their intended purpose, they are not necessary for solar observing, as you do not need a computer to find the Sun!

Whatever type of telescope mount you choose, make sure it is steady enough. A good test of a mount's stability is to give the telescope tripod a sharp knock and time how long the vibrations in the image take to die out when a high magnification is being used. If the vibrations take longer than a few seconds to settle down, you may wish to consider upgrading to a better-quality mount. Similarly, check that the slow motions – manual or electronic – turn smoothly in all directions.

Observing the Sun by Projection

Solar projection is traditionally the most popular method of viewing the Sun's image. Reduced to its basics, projection consists of holding a simple white screen, such as a sheet of white paper, 30 centimetres (12 inches) or more behind the eyepiece of the telescope and focusing the telescope so that a sharp image of the Sun is formed on the paper. The image on the paper is formed in a similar manner to that produced on a film by a camera lens. In my opinion, projection remains much the best method for observing the Sun with a refractor. It is less convenient with a reflector, however, and it is totally unsuitable for Maksutov or Schmidt–Cassegrain telescopes. If you have one of these telescopes, you need to use one of the alternative methods described below.

As an experiment, you can project the Sun's image quite simply by holding a sheet of A4-sized white paper between 30 and 50 cm behind the eyepiece of your telescope (Figure 2.3). Shade the image from the ambient sunlight by fitting a sheet of card around the telescope tube. A low-power eyepiece will project an image of the Sun a few centimetres across – easily large enough to show the principal sunspots visible on the disc. Eyepieces of higher magnification will show more detail in the sunspots but give a fainter image. Moving the paper further behind the eyepiece also gives a larger but fainter image.

Figure 2.3. Projecting the Sun's image onto paper for safe solar viewing. A card has been fitted over the eyepiece end of the telescope to shade the image from direct sunlight.

Some small refractors come with a ready-made solar projection screen. This usually takes the form of a small sheet of white-painted metal attached to a long stalk, the other end of which is clamped to the eyepiece end of the telescope. This setup will also show the main sunspots. However, while commercial screens such as this and the simple method above are fine for experimentation or casual viewing, they are not adequate for serious observing. The main reason is lack of contrast. Even with a sunshade, the daylight makes the image much fainter than it should be and can all but wash out delicate solar details. Also, you cannot make accurate observations with a hand-held or commercial projection screen.

Fortunately, it is possible to shade the Sun's image very effectively using a box that holds the screen to the telescope while keeping out most of the daylight. Projection boxes are not available commercially, but it is quite easy to make one yourself. The entire box is covered in, except for part of one side, which is kept open to allow the Sun's image to be seen. Attaching the box to the telescope leaves both your hands free to make notes, drawings and adjustments. The telescope eyepiece protrudes into the box through a hole in the eyepiece end of the box just big enough to fit over the eyepiece or drawtube and projects the image onto a white screen at the far end of the box. In order not to upset the balance of the telescope tube, you need to make the box from a very light material. Balsa wood is excellent, as it is light but strong and can easily be cut and manipulated using ordinary hand tools. You can obtain balsa wood from toy shops, model shops or suppliers to the packaging industry.

How you build your box is up to you, as amateurs have achieved excellent results with many different designs. The diagram (Figure 2.4) and the photograph (Figure 2.5) show the design for my own projection box. I made the top, bottom and side pieces from 1 cm ($\frac{3}{8}$ inch) thick balsa wood bought from a toy shop and covered in the front face with a sheet of thick black paper from an art shop. I also lined the interior of the box with black paper to reduce internal reflections, which can spoil the contrast of the image. The box is square in its dimension perpendicular to the telescope tube, so that I could join the pieces together easily using balsa wood framing of square cross-section, 12.5 mm ($\frac{1}{2}$ inch) on a side. However, I made the sides of the box so that they tapered down to 75 mm (3 inches) at the

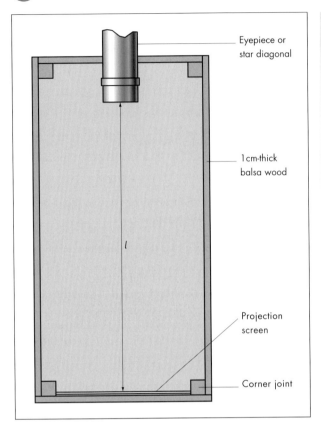

Eyepiece or
star diagonal

1 cm-thick
balsa wood

l

Projection
screen

Corner joint

Figure 2.4. Diagram showing construction of the author's projection box. The length *l* determines the diameter of the projected image.

eyepiece end, as otherwise the front of the box would collide with parts of the tube. I glued the joints using ordinary wood adhesive from a local hardware store.

The most important dimension to establish when building a projection box is its length. Using an eyepiece of a given magnification, moving the screen further behind the eyepiece increases the size of the projected image. To do serious solar work you need an eyepiece which will show the whole solar disc comfortably within the field of view. Avoid using an eyepiece which shows the limb of the Sun's disc very close to the edge of the field, as the view at the edge of the field is distorted in most eyepieces. On the other hand, using too low a power gives a very long projection distance for a good-sized disc and so an inconveniently long projection box. I have found that an eyepiece giving a magnification of between 50× and 60× gives the best results. The standard diameter used by amateur astronomers for a projected solar image is 152 mm (6 inches), although if you use a very small telescope you may find a 100 mm (4 inch) disc easier to

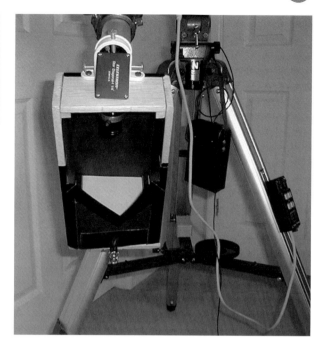

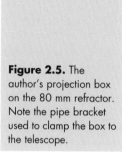

Figure 2.5. The author's projection box on the 80 mm refractor. Note the pipe bracket used to clamp the box to the telescope.

work with. To find the correct length for a projection box on your own telescope, draw a circle of your chosen disc size on a sheet of plain paper. Insert your choice of projection eyepiece in the telescope and measure the distance behind the eyepiece you have to hold the paper in order for the Sun's image to exactly fill the circle. The final length for your projection box is this distance (marked l in Figure 2.4) plus the distance by which the eyepiece protrudes into the box. The precise value of the latter distance is determined by how you clamp the box to the telescope (see below). Because the Earth's orbit is slightly elliptical, the Sun's apparent diameter varies by about an arc minute over the course of the year. It reaches its largest apparent diameter in January and the smallest in July. This means that the precise projection distance varies with the time of year. However, the variation is never more than a centimetre or so and it should not be difficult to adjust the position of the box or the eyepiece by this amount.

The screen end of the box can be square. When determining its size, allow for 12.5 mm ($\frac{1}{2}$ inch) of clearance around the solar disc, plus the thickness of the square corner joints (another 25 mm or 1 inch in total) and the thickness of the sides of the box (2 cm or $\frac{3}{4}$ inch). So if you have chosen to use a projected image

4 inches in diameter, the screen end of your box needs to be $6\frac{1}{2}$ inches square, while if you use a 6 inch solar disc it needs to be $8\frac{1}{2}$ inches square.

How you attach the box to the telescope depends on the characteristics of your instrument. Ideally, the box should slide over the drawtube and be clamped either with a screw or by making the box fit very tightly over the drawtube (e.g. by lining the eyepiece hole with felt). However, some telescopes may not reach focus without using a star diagonal. In this case, you need to slide the box over the eyepiece end of the star diagonal, so that its long dimension is at right angles to the telescope tube. Clamping the box rigidly to the telescope requires some ingenuity if you use a star diagonal. I employed a pipe bracket (again obtained from a hardware store) attached to the eyepiece end of the box, with a bolt passing through it and so clamping the box tightly to the telescope (see Figure 2.5).

To obtain the best projected images you should use a good quality, thick paper. Many amateurs over the years have used "Bristol Board", a very smooth and white paper available from art shops. I, too, have found this to give good results, but I have found that thick, smooth cartridge paper with a slightly creamy tint shows sunspot detail even better. Faculae, though, are more prominent on Bristol Board. In my projection box I employ a sheet of Bristol Board and a sheet of cartridge paper pasted together, so that I can flip between the two surfaces to see the two types of solar features to their best advantage.

A few amateurs, particularly those with observatories or who do their observing from indoors, mount their projection screens entirely separate from the telescope, on a tripod or other support. I do not recommend this practice for accurate work, as even if the telescope has a motor drive the Sun is always moving relative to the screen. However, observing from indoors out of an open window has much to be said for it, as your surroundings are generally darker and so the image contrast is increased still further. It is only possible in the warmer months, though, as in winter the mixture of warm and cold air causes severe air currents that ruin the image.

You may ask why, when a number of solar filters have been available commercially for years and have been tried and tested to be safe, I still recommend the old-fashioned projection method. The reason is that projection has several distinct advantages over filters. The first is that it is completely safe. Filters can be damaged and even small scratches or pinholes in their delicate metallic coatings can allow dangerous radiation to pass through. It is also possible for filters to be blown off the telescope by gusts of wind or fall off if not properly secured – with disastrous consequences for the observer's eyesight. Projection is also much the cheapest method of viewing the Sun. The projection box described above cost me only about £5 ($8) to build, whereas some filters for the same size of telescope cost more than ten times this amount. Thirdly, projection allows the Sun to be observed by more than one person at the same time – a unique advantage over any other astronomical object, where people have to take turns at the eyepiece and the telescope has to be adjusted each time. This makes projection ideal for teaching purposes and group viewing at astronomy clubs. Finally, for the serious solar observer, a projected image makes it much easier to plot the positions of sunspots and to make accurate counts of sunspot numbers.

Projection does, however, have one major problem – heat. Because the telescope is unshielded, heat builds up inside the telescope tube when the instrument is pointed at the Sun. In particular, the eyepiece, because it lies near the telescope's prime focus, can become very hot and so great care must be taken when choosing eyepieces for solar projection. The advice of experienced solar observers for many years has been never to use any eyepieces except the very simple types such as the Huygens and the Ramsden, and avoid altogether the more complex types, such as Orthoscopics and Plössls. The reason is that the more complex eyepieces contain several lens elements bonded together with optical cement. The intense solar heat can melt the cement and so damage these eyepieces. Huygens and Ramsden eyepieces each contain just two completely separate lenses, with no cement, and so they can withstand prolonged exposure to the Sun.

Nowadays, however, this advice requires serious qualification. I personally use Orthoscopic eyepieces for all my solar work with the 80 mm refractor, because they give much better images than the simpler types. But I always take the precaution of observing the Sun when it is less than 30 degrees above the horizon, when its radiation is much less intense. For reasons described in the next chapter, it is never a good idea to observe the Sun when it is high in the sky anyway. I also cover the telescope's object glass to let the eyepiece cool after half an hour of continuous solar observation. I have used these eyepieces for several years with both 60 mm and 80 mm refractors and they have never suffered any damage due to heat. Moreover, Huygens and Ramsden eyepieces are becoming difficult to get hold of nowadays and those that are available are often made of plastic, which will quickly melt when exposed to the Sun. Provided that the above precautions are taken, I believe that Orthoscopic eyepieces are quite suitable for solar projection with 80 mm or smaller refractors. If you use a reflector or a larger refractor, however, you need to reduce the telescope aperture to 80 mm or smaller, using an off-axis mask as described above. Do not use "exotic" wide-field eyepieces, such as the Ultra Wide Angle or Nagler varieties for solar projection. These eyepieces were designed for deep-sky observing and their wide fields are not necessary for solar work. They would also cost a lot of money to replace if they were damaged by the Sun's heat.

Projection works excellently with refractors and while it is not so convenient with reflectors, it is usually possible to rig up some sort of projection arrangement. You may have to be ingenious and attach some heavy object, such as a small tin or bag filled with nails, to the bottom end of the tube to balance out the weight of the projection box. If you have a Maksutov or Schmidt–Cassegrain telescope, however, you have to employ a different method, and this is where difficulties can begin.

Solar Filters

The subject of filters is not easy, as many filters are not safe for solar viewing. One of the most dangerous types of filter is that designed to screw into the eyepiece. These eyepiece filters take the form of a piece of dark glass encased in a metal ring and marked "SUN". They sometimes accompany the very cheap, small refractors sold in high-street stores and are another reason for avoiding these telescopes. These filters are dangerous partly because they pass unsafe levels of radiation, but also because their proximity to the telescope's focus can cause them to crack without warning, instantly enabling the Sun's heat and light to pass unchecked through to the eye. Fortunately, these filters are less common than they used to be. Never be tempted to use one, and never use *any* filter at the eyepiece end of the telescope.

Only filters mounted at the *front* of the telescope are safe, because they are not subjected to the focused heat inside the telescope and only receive the normal daytime solar heat. Such filters are known as *aperture filters*. However, not all aperture filters are safe or suitable for solar observing through a telescope. In particular, never use home-made filters. They may *appear* to reduce the Sun's light and give an image faint enough not to dazzle your eyes, but they let through far too much invisible radiation and so damage your eyes before you realise it. Examples of dangerous home-made filters include sunglasses, polarising and Neutral Density photographic filters, CDs, CD-ROMs, computer floppy disks and food wrappers, including those with a silvery appearance that appear to reflect the Sun's light.

Some older sources recommend using black-and-white photographic film as a solar filter. The film is unravelled and exposed to normal daylight before being fully developed, so making it turn as black as possible. The silver halide crystals in the film cut out the infrared radiation. I do not recommend this for the modern solar observer, as film might still not block enough infrared for use with a telescope and in any case, not all black-and-white films contain silver halide nowadays. This advice was intended in the past for those observing a solar eclipse with the naked eye. The optical quality of photographic film is also not nearly good enough for the telescopic observer. Even with the naked eye, the Sun has a noticeable "fuzz" around it when viewed through this type of filter. Colour film of any type should NEVER be used as a solar filter, for naked eye or telescopic observing. Although it absorbs visible light well, it is nearly transparent to infrared radiation.

Another traditional filter, again intended for naked-eye viewing but worth mentioning here, is smoked glass – a piece of glass coated in soot by holding it over a candle flame. This is very risky because it is easy to coat the glass in too thin a layer of soot and it is very difficult to get an even layer. There is therefore the risk that the amount of filtration might not be enough. The soot layer is also very easy to smudge and damage. In any case, with so many good filters available on the market nowadays, there is no need for the modern amateur to resort to such a primitive device.

The only safe solar filters are *filters made specifically for solar observing through a telescope*. These fall into two types. The cheapest is a sheet of polyester film, often known by the trade name Mylar, coated with a thin layer of aluminium. Mylar is available from astronomical suppliers either in mounts designed to fit a particular make and model of telescope or as a sheet from which the amateur can cut a portion to place in a home-made mount. Never confuse silver food wrappers or "space blankets" with Mylar filters – they are not produced to optical standards and let through far too much infrared radiation. Some of these materials are not even Mylar, but just ordinary plastic coated with silver paint. Before buying Mylar – or, indeed, any solar filter – check with an experienced amateur that it is safe and of good quality. When you take delivery of the filter, hold it up to the light and see if there are any scratches or pinholes which let the light through. If it has any such defects, do not use it, as they can let through dangerous radiation. Check with the supplier that the Mylar is coated on both sides; this offers extra protection against scratches or pinholes. When you receive a mounted Mylar filter, you may be surprised that the material is slightly loose and wrinkly in its mount, but this has no effect on its optical performance. In fact, if you choose to mount the Mylar yourself, never pull it taught, as this can damage the surface.

When properly mounted up and attached to the front of the telescope, a Mylar filter gives a solar image with a strong blue tint. This is because aluminium passes blue light more efficiently than red. It also scatters light slightly, making the sky background look very faintly blue. Both these effects can be reduced somewhat by threading a light yellow filter (also available from astronomical suppliers) into the eyepiece, *in addition to* the Mylar mounted over the aperture. This gives the Sun a more realistic colour and can also increase the sharpness and contrast of the image. In the late 1990s, the German company Baader Planetarium introduced a variation on Mylar known as AstroSolar Safety Film. To outward appearance it resembles Mylar, but the manufacturer claims the material has undergone special treatment and that its radiation transmittance has been precisely measured. It certainly gives results far superior to conventional Mylar filters. The blue colour is hardly noticeable at all – indeed, the Sun looks almost white – and the sky background is almost totally black. The solar image is very crisp and sunspots show up in beautiful detail. Some companies supply mounted filters made from AstroSolar, while the material is also available in rolls or A4-sized sheets with instructions on how to make your own mount (Figure 2.6). Buying the material unmounted is very good value for money, as you can share the cost of sheets or rolls with other

Figure 2.6. Aperture filter made from Baader AstroSolar Safety Film in a home-made mount, fitted to the author's 80 mm refractor.

interested members of your local astronomy club, but be sure to follow the manufacturer's instructions for safely mounting the material. AstroSolar is also available coated to a lighter density for photographic use, but this is not safe for visual observing. I will discuss solar photography with filters in Chapters 7 and 8.

More expensive are the glass solar filters supplied by several astronomical instrument makers. These consist of a disc of glass polished on both sides to optical quality and coated on one side with a layer of stainless steel (sometimes known as "Inconel"), chromium or an alloy. Most glass filters come in mounts made for a specific telescope. They are sometimes available in unmounted format, but the small saving in price is not worthwhile, because glass filters are more difficult to mount securely than Mylar filters. Again, if you purchase such a filter, you should check it carefully for defects. Unlike Mylar, glass filters give a solar image coloured yellow or orange, due to the different transmission characteristics of the coating, and so there is no need to use a secondary yellow filter. The sky background is very dark and sunspots show up with excellent contrast. Glass photographic filters are also available, but again these are not suitable for visual observing.

Glass filters are more durable than Mylar filters. Although they can cost up to ten times as much as Mylar filters, it may be worth considering purchasing a glass filter if you use a catadioptric telescope and so have to use a filter all the time. Both types of filter, however, should be stored carefully when not in use, as prolonged exposure to air eventually degrades them. A glass filter can be stored in the box or plastic sleeve it comes in, while an airtight plastic box of the sort intended for storing foods makes a good container for a Mylar filter.

Once again, always check out a solar filter carefully before using it. Sometimes even aperture filters advertised as being suitable for solar observing turn out to be unsafe. A classic case happened in the UK during the early 1990s, when advertisements for solar filters appeared in a British astronomy magazine. The filters were found to be pieces of coloured plastic and totally transparent to infrared radiation. One experienced solar observer dramatically demonstrated this at an astronomy meeting by pointing a television remote-control unit through the filter and switching on the TV! Enough of the radiation from the unit's infrared beam was getting through the filter to the TV. Only buy filters from reputable, established astronomical suppliers and even then get an experienced observer to check your filter before using it.

Other Observing Methods

Before the advent of safe aperture filters, there were two alternatives to solar projection. The first was the "Herschel Wedge", sometimes called a Sun diagonal, and invented by Sir William Herschel. This resembles an ordinary star diagonal as supplied with refracting and catadioptric telescopes and takes the form of a prism whose housing has a hole at the rear. The prism reflects a small percentage of the Sun's light up to the eyepiece while the rest of the light passes through the prism and out of the hole at the back. I do not recommend these devices, however, as the amount of radiation passed is still dangerously high. It is also very easy to burn yourself with the unchecked solar heat emerging from the back of the prism.

Another alternative is the dedicated solar telescope. Several designs for these have been tried, although the best-known is an ordinary Newtonian reflector with an unsilvered mirror. This design works in the same way as the Herschel Wedge, the mirror reflecting a small amount of light back to the eyepiece and passing most of it out of the back of the tube. Such a telescope still requires a secondary filter at the eyepiece and, unless an aluminised mirror were substituted for the unsilvered one, it could not be used for anything but solar observing.

Many astronomy books recommend using a welder's glass as a solar filter. Provided that it is of shade number 14 or higher (i.e. denser), a welder's glass is indeed safe for viewing the Sun with the naked eye, as these devices are designed for preventing harmful radiation from reaching the eye. I do not recommend it as a telescopic filter, however, as its infrared blockage may still not be enough for the far greater power of a telescope. Also, it is not made to optical standards and so would not provide as sharp a view as a conventional solar filter. Additionally, a welder's glass gives the Sun a lurid shade of green.

Observing the Sun with the Naked Eye and Binoculars

For observing the Sun's larger spots you do not even need a telescope. Very large sunspot groups can sometimes be seen with the naked eye, provided that it is protected by a safe Mylar or glass solar filter as described above. A welder's glass of the appropriate shade number is also suitable as a naked eye solar filter. Never be tempted to examine the Sun through mist or when it is low in the sky – you can never be sure that enough radiation is being blocked out. When you see a large sunspot using your telescope it is interesting to see if it is visible with the protected naked eye and, as I will describe later on, some amateurs carry out systematic counts of naked eye sunspots.

Many astronomy books recommend using a pair of binoculars if you cannot afford a good-quality telescope. This is sound advice and it applies to the Sun as well. Binoculars are especially suited to projecting the Sun indoors: simply mount them on a camera tripod and let them throw an image of the Sun on to a white screen mounted one or two metres away. Closing the curtains around the binoculars will darken the room and so increase the contrast of the projected image. Be careful not to let the unfiltered sunlight coming through the binoculars glint in your eye and always keep children or uninformed people under supervision. Also, do not leave binoculars pointed at the Sun for too long, as the concentrated solar heat inside the instrument may damage the cement holding the prisms in place. You can also use binoculars with a pair of Mylar filters, provided that you mount the filters safely as described above. Some companies supply Mylar filters in pairs mounted to fit larger binoculars. If you have smaller binoculars you will need to buy a sheet of Mylar and make your own mounts.

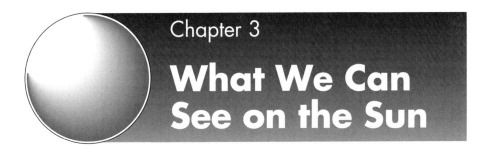

Chapter 3

What We Can See on the Sun

Setting up for Solar Observing

Before we discuss what can be seen on the Sun's image we need to know what are the best conditions for observing the Sun. Our atmosphere interferes with solar observing in two important respects. First, it produces clouds which obscure the Sun or allow us only a poor view of it. Secondly there is the constant rippling and blurring of the solar image, which is caused by the fact that our atmosphere is in constant motion above our heads. Both these atmospheric factors are a nuisance to the amateur solar observer and are the main reason why professional astronomers site their solar observatories on the tops of high mountains or on spacecraft. However, with a little planning and thought about the causes of these problems the amateur can sometimes get round them to some extent.

When to Observe the Sun

Before we discuss the atmospheric difficulties there is the question of getting access to the Sun in the first place. This is not as obvious as it sounds, because the Sun's position in the sky varies greatly according to the time of day and time of year. As seen from temperate latitudes, the Sun in summer rises on the north-eastern horizon and sets in the north-west, passing very high in the south at midday. In winter it rises in the south-east and sets in the south-west, and even at noon it is still quite low in the sky. Overall, I have found that the best time of year for observing the Sun is summer, simply because the Sun is then more accessible. Between May and August the Sun is visible until about 9 p.m., Summer Time (assuming your local Summer Time is one hour ahead of local time) and this allows us to observe it during weekdays in the evenings after our work and daily activities are over. In winter solar observing tends to be a weekends-only pursuit for most of us.

Oddly enough, the best time of day in which to observe the Sun is *not* midday, when the Sun is at its highest. This is because of the phenomenon of convection, in which the strong heat of the midday Sun warms up the ground, which in turn re-radiates this heat into the air, causing bubbles of warm air to rise and "stir up" the atmosphere above. Convection is a major cause of the rippling and blurring of the Sun's image mentioned above. Astronomers call the amount of rippling and blurring the "seeing". In good seeing the rippling is only minor and the Sun's image can sometimes appear rock steady for a few seconds at a time. Bad seeing is characterised by a constant, large-scale rippling of the image, with the Sun's limb appearing to "boil" and fine details in sunspots and other solar features become invisible. Convection is at its worst around midday, when solar heating is strongest, and for this reason it is wise to avoid observing at this time if you can. This is another reason for choosing summer evenings as a good observing time. By evening the Sun is lower in the sky, its heating effect on the ground is less and so convection is much reduced. The early morning is also a good time to observe, because at that time the ground is still cool after the night and the Sun has not yet heated up the ground strongly. If you can observe at this time, you may get even steadier views and it is well worth trying. However, for practical reasons the early morning is often an inconvenient time for many people, and good observations are harder to make if you are pressed for time. Many people, I feel, will still prefer to observe in the evening. I have found the steadiest conditions to be between an hour and three hours before sunset. Before this time interval the summer Sun is high enough for convection to be a problem. Later on the Sun is low in the sky and its light shines through a long path in the Earth's atmosphere, and the steadiness of the image declines again.

In winter the solar observer has different problems, as between November and February the Sun never rises high enough to cause much convection. At this time of year observing the Sun at or around midday makes good sense, because the Sun is very low in the early morning and afternoon. But if you live in an urban area, the Sun's image can be affected by convection from a different source – the rooftops of houses warmed by central heating. If you observe the Sun when it is low over a rooftop the seeing can often be ruinous.

The weather is a major factor in both the accessibility of the Sun and the seeing conditions. Unfortunately for observers in many parts of the world, this is not easy to predict, but some general common-sense rules can be applied. I and many other amateur astronomers have found that the best seeing usually occurs during high-pressure conditions. In summer these often cause hot, dry and sunny spells and their winter equivalents cause cold, frosty conditions. A strong area of high pressure can sometimes keep other weather systems out and so such conditions can persist for many days at a stretch. For the solar observer, it is in such conditions that the atmosphere is at its most stable and seeing is consistently good. Occasionally, seeing conditions can be really superb, the Sun appearing rock steady and details in sunspots and the surface granulation visible down to the resolution limit of the telescope. I have sometimes found such exquisite seeing to occur on a sultry and slightly hazy summer evening, when there is little or no wind and the sunlight appears dim and yellow. Conditions such as this are hard to come by in winter, although seeing can be very good on a frosty day, provided you are not looking over rooftops. In cold weather it is also a good idea to leave the instrument outside for an hour or so to reach the same temperature as the outside air, as warm air currents from a telescope brought out of a warm room can also spoil the image.

Where to Observe the Sun

Your choice of observing site can also have an effect on the seeing, as some atmospheric turbulence is caused by local factors. Most of us have very little freedom as to where to observe on a regular basis, as our only site is where we happen to live. However, some locations within a house or garden can provide steadier views than others. I have already mentioned above that poor seeing results from observing the Sun when it is low over rooftops.

Even in summer, rooftops absorb a lot of solar heat during the day and they constantly re-radiate it back into the atmosphere. The seeing is always far better above trees or grass. For the same reason, seeing can be mediocre in areas containing large stretches of concrete or tarmac. If you can, try to observe from a lawn or park rather than, say, a driveway or car park – the improvement in seeing is usually quite noticeable. This is one area where a portable telescope scores over a permanently sited one, as it is easy to move a portable instrument to a better site at short notice.

I mentioned briefly in Chapter 2 that solar observing does not necessarily have to be done outside. Indeed, between May and August I do practically all my solar observing from out of a first floor window, which conveniently faces west-northwest, where the Sun is on summer evenings. Many astronomy books advise strongly against observing the night sky out of a window, and for good reason. At night there is always a large temperature difference between the inside and outside air, even in summer, and opening a window causes warm air to rush out into the colder exterior, resulting in disastrous seeing through the telescope. During the daytime, however, things are different. In the summer months there is often little or no temperature difference, and views out of a window are every bit as good as they are from outside. If you use the projection method, observing out of a window greatly improves the contrast of the solar image, as the amount of ambient light is greatly reduced, especially if you draw the curtains around the telescope. Observing from indoors has the added advantage of comfort for the observer, as you are shielded from the heat of the summer Sun. I do my summer solar observing from an armchair pulled up in front of the projection screen – giving a new meaning to the phrase "armchair astronomy"! Finally, if you keep your telescope set up indoors, observing out of a window allows access to the Sun at a moment's notice – essential if you live in a cloudy climate. During the winter months, though, it is still necessary to observe from outdoors, as any heating inside a room creates enough of a temperature difference to ruin the Sun's image.

Heat within the Telescope

A final source of poor seeing can come from within the telescope itself. Any telescope that is used to project the

Sun's image is a potential solar furnace, because it allows the Sun's heat to enter the optical system unchecked by any filter. The focused heat inside the telescope heats up the air within the tube, causing air currents to circulate. If you observe the Sun's projected image on a warm day over a period of about half an hour or so, you will find that the image quality gradually deteriorates, regardless of how favourable the weather or your observing site is. The main remedy for this problem is not to observe the Sun for too long continuously, but cover the object glass for a few minutes every quarter of an hour or so to let the telescope cool down. Heat build-up in the telescope is another reason for avoiding observing when the Sun is at a high altitude, because it is always less of a problem when the Sun is lower in the sky. Another possible solution to the heat problem is to stop down the telescope's aperture. My 80 mm refractor, like many small refractors nowadays, contains a small secondary cap in its objective lens cover. When the small cap is removed and the main cover left in place, the telescope's aperture is effectively reduced to about 40 mm. I use this arrangement when making my initial reconnaissance of the Sun's disc at low power, and I only remove the main cap when I examine details of the spots at high magnification. In this way I ensure that for some of the time the amount of heat entering the telescope is reduced.

Aiming the Telescope

Now that we have decided on the best time and place for solar observing, there is the question of centring the Sun's image in the telescope. Because you cannot look at the Sun through a telescope, you have to find it by some indirect means. Finding the Sun is a problem whether you use projection or a filter, because while an aperture filter allows you to view the Sun safely, you still have to aim the telescope at it. You cannot use the finderscope, which is every bit as dangerous to look at the Sun through as the main telescope. Neither should you sight the telescope on the Sun by looking along the edge of the tube, as this is also dangerous. You could locate the Sun using your telescope's setting circles and an ephemeris such as the BAA *Handbook*, and most computerised telescopes include the Sun in their databases. However, this is not an option unless your telescope is permanently sited, as you cannot polar-align your telescope during daylight.

A traditional and safe method of locating the Sun is by using the telescope's shadow. When a telescope is pointed in the direction of the Sun its shadow is at its smallest. When the telescope's shadow is as small as you can get it, look at the projected image or, if you use a filter, in the eyepiece. You may need to sweep the telescope back and forth slightly to bring the Sun into the field of view. Using a low-power eyepiece makes it easier to find the Sun this way.

Safe as the shadow method is, it also involves a bit of trial and error and can be tedious every time you want to observe the Sun. If you live in a damp climate, such as Britain or the north-eastern USA, it is often necessary to find the Sun quickly before clouds roll in, and finding the Sun by the shadow method can waste several minutes of precious observing time. Some amateur astronomers, therefore, have made their own "solar finderscopes", which employ the projection method as in the main telescope. These take the form of two small pieces of wood or cardboard, one mounted in front of the other along the length of the tube and spaced a few centimetres apart. The front piece has a small hole about 1 or 2 millimetres in diameter punched in its centre. This acts rather like a pinhole camera, throwing a tiny image of the Sun onto the rear piece, which has a cross or mark inscribed on one side. When the miniature solar image falls on the mark the telescope is exactly pointed at the Sun and the solar image should appear in the eyepiece or projection box. Of course, like a conventional finderscope, such a device needs to be accurately aligned with the main telescope in order to work.

Some years ago I made myself a variation on the solar finderscope. I then used a 60 mm refractor as a solar telescope, but I did not use it for any other type of observing, as I also had a larger-aperture instrument. I therefore removed the telescope's conventional finderscope, which was of no use for observing the Sun (Figures 3.1 and 3.2). The finderscope's mount was composed of two metal rings, one in front of the other. To each ring I glued a small disc of Bristol Board (the same thin white card used for projecting the Sun's image), just big enough to cover the ring. When the glue had set I punched a small hole in the front disc. The next time I observed the Sun I found the Sun in the telescope using the shadow method, and when its image was accurately centred in the projection box I carefully marked on the rear disc the spot on which the tiny pinhole solar image fell. On every observing session after that I could find the Sun instantly, enabling me to spend more of my solar observing time observing.

A final point about finderscopes. When you are using a telescope to observe the Sun, make sure both lens caps are securely placed on the finder – or, best of all, remove the finderscope altogether. Lens caps can be removed by curious children and in any case, it is all too easy to look through a finderscope by accident. You have been warned!

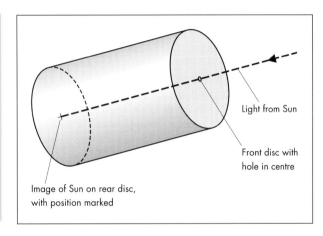

Light from Sun

Front disc with hole in centre

Image of Sun on rear disc, with position marked

Figure 3.1. Diagram showing the principle of a solar finderscope.

Figure 3.2. Solar finderscope on the author's 80 mm refractor. This replaces the "normal" finderscope, which has been removed as a safety precaution.

The Sun's Surface

When observing the Sun, it is a good idea to always start with a fairly low magnification, so that the entire disc of the Sun is visible on the projection screen or in the telescopic field. Not only does this make finding the Sun easier, but it also enables you to see the entire Sun at a glance and identify any interesting areas. You can then hone in on an interesting sunspot group using a higher power. I use two eyepieces with my 80 mm refractor: an 18 mm Orthoscopic for viewing the whole disc, and a 7 mm Orthoscopic for detailed views of solar features.

Usually the most obvious features to be seen on the Sun are the sunspots, but before we discuss these let us look at a couple of other general characteristics of the Sun.

Granulation

When viewed at a low magnification, the solar surface appears smooth and uniform. However, when a medium or high power is employed, and providing the atmosphere is reasonably steady, the Sun shows a faint granular texture, known as the solar *granulation*. The granules are bright polygon-shaped features separated by darker lanes. Each granule is the top of a column of gas rising up from the convective layer inside the Sun discussed in Chapter 1. The dark lanes are material falling back into the Sun. In a telescope the granules appear tiny, only about 1.5 arc seconds across and constantly shimmering due to turbulence in our atmosphere. But it gives some indication of the vast size of the Sun when we realise that each granule is about 1,100 kilometres (700 miles) across – larger than the whole of France! Granules are very short-lived, lasting only about 20 minutes on average, and they are constantly fragmenting and re-grouping with other granules. The granulation is not easy to observe in a telescope, partly due to the effects of our atmosphere but also because it is quite subtle and indistinct. These factors make it difficult or impossible to follow the life and death of an individual granule. You may find it easier to see the granulation by slowly moving the telescope back and forth using the slow motion controls. The most spectacular views of the granulation are to be had during occasions of exceptionally good seeing, such as the humid summer evenings or the frosty winter days mentioned above. In fact, granulation is a good rough measure of the quality of the seeing. If the seeing is very bad, granulation is difficult or impossible to see at all. On most days it is visible to some extent, but turbulence nearly always degrades our view of it.

Limb Darkening

Switching back to a low-power eyepiece, you may notice another characteristic of the Sun's visible surface: it appears brighter at the centre of the disc than at the edge, or limb. This *limb darkening* is caused by the fact that the photosphere is a thin layer and that its temperature decreases with height, from over 6,000°K at the bottom to just 4,400°K at the top. When we look at the centre of the solar disc, we are looking at the photosphere from directly above it, and so we are looking at the hot inner part of the photosphere through only a thin layer of cooler gas. But because the Sun is a sphere, when we look near the limb we are looking at the higher, cooler layers of gas, which appear darker. Limb darkening has one important effect on solar observing. It allows us to see the *faculae*, brighter patches on the Sun which often appear around or near sunspot groups. Faculae are usually invisible when they are near the Sun's centre, where they are masked by the intense brightness of the surrounding photosphere.

Sunspots

Sunspots are not always present, and within a year or so of solar minimum there may be periods of days or even weeks when there are no sunspots visible. Outside this time, however, there are usually at least a few spots visible on the Sun's disc.

The first thing to notice about sunspots is that they come in all shapes and sizes, from the tiniest dots, barely perceptible even when a high magnification is used, to huge, complex blemishes, showing considerable detail even at low power. The smallest spots visible in amateur-sized telescopes have diameters of around 1,000 kilometres (600 miles), which at the distance of the Sun translates into an angular diameter of 1 second of arc ($1''$) – the resolution limit of a small telescope. A typical small to medium-sized spot may have a diameter comparable to that of the Earth (12,756 km, about 7,800 miles). The very largest sunspot groups can grow to 100,000 km (60,000 miles) or more in their longest dimension. Large spot groups, however, are not as common as small ones and the very biggest ones generally appear only near solar maximum.

Sunspots show obvious changes from one day to the next, in terms of their position, their shape and their internal structure. In fact, the thought of what might be new on the Sun and what might have changed since yesterday drives the keen solar observer to observe the Sun on a daily basis. From one day to the next, the Sun and its spots are never the same.

The main positional change you can see from day to day is that the sunspots move from east to west on the Sun's disc due to the Sun's rotation. A typical sunspot takes about two weeks to move from the eastern limb to the western limb of the solar disc. As I explained in Chapter 1, the Sun's rotation is differential, i.e. it rotates faster at the equator than it does at the poles. Near the equator the Sun's rotation period is 25 days, but as we look nearer the poles this increases to 30 days. Long-term observations of sunspots show that spots at high latitudes get "left behind" by those nearer the equator. The effect of differential rotation on sunspot positions is more subtle than the general day-to-day solar rotation, but it is possible to observe it using photography or precise plotting of sunspot positions.

Another interesting thing to notice about sunspot positions is that sunspots are never to be found anywhere near the northern or southern limbs of the Sun. In fact, sunspots are confined to two parallel bands surrounding the equator. Sunspots are not usually seen exactly at the equator, but they can often occur just a few degrees of solar latitude away from it. The main "belts" of sunspot activity lie at between 5° and 25° solar latitude – although, as I shall explain below, sunspot latitudes vary with the solar cycle. Sunspots start to thin out at 30° or so and beyond 40° they are quite rare.

Sunspots also show considerable daily changes in their shape and structure. A small, single spot seen one day might be a pair of spots the next. A day or two later on it might have grown and contain several more spots. By the end of a week it may have grown to become quite a large and complex group. As a group grows, it tends to flatten out with respect to solar latitude. For example, a small group may be inclined at a consider-able angle to the solar equator to begin with, but as it gets larger this inclination decreases. Very large groups

Figure 3.3. Whole-disc photograph of the Sun, showing some simple sunspots, composed of dark central umbrae surrounded by lighter penumbrae.

tend to be long and thin, covering many degrees of longitude but often just a few degrees of latitude.

Before we look in detail at sunspots and their characteristics, let us consider a typical sunspot such as one of those shown in Figure 3.3. The photograph shows some simple examples, but such sunspots are very common and it illustrates features common to the majority of sunspots. A sunspot contains two sections: a dark central part, known as the *umbra* and a somewhat less dark surrounding region, or *penumbra*. The relative sizes of the two sections vary from sunspot to sunspot and sometimes a sunspot group contains mostly penumbra and very little umbra. This is often the case with a large group that has passed the peak of its activity and has "decayed". At first glance the umbra appears very dark, but it is not really black. As I noted in Chapter 1, sunspots only appear dark by comparison with the even hotter surrounding photosphere. Umbrae are, in fact, greyish-brown in colour, although in refractors this is sometimes masked by the effect of chromatic aberration, which can give them a purplish tinge. At first glance, the surrounding penumbra appears smooth and grey through a small telescope. However, if you are using an 80 mm or larger telescope and the seeing is very steady, you may notice some fine structure within it. A sunspot penumbra is composed of myriads of extremely fine dark streaks radiating outwards from the umbra. These streaks are separated by lighter streaks known as *penumbral filaments*. These features are only a fraction of an arc second across and so in theory should be invisible in small telescopes. However, high-contrast features with a long, thin shape are often visible beyond the resolution limit of a telescope. The same is true of linear valleys on the surface of the Moon, some of which are only a few hundred metres across. Craters of the same diameter are not visible in the same size of telescope.

The type of sunspot I have described here is known as a "symmetrical" spot, because it has a circular outline when it is near the centre of the solar disc. As it approaches the limb, however, the effect of perspective causes it to look elliptical. As long ago as 1769 the astronomer Alexander Wilson of Glasgow University noticed that in a sunspot near the western limb the umbra was displaced towards the centre of the disc; in other words, the penumbra on the side of the spot nearest the centre of the disc was noticeably narrower than that on the side facing the limb. The phenomenon became more noticeable as the spot came nearer the limb and was still visible when the same spot reappeared at the eastern limb a fortnight later. The effect became known as the "Wilson Effect" and it led Wilson and other astronomers to believe that sunspots were depressions in the solar surface, rather like craters on the Moon. We now know, of course, that the Sun's surface is a gas and that no craters could ever exist there, but Wilson's conclusions were partly true. Astronomers now believe that the density of the gas inside a sunspot, and especially the umbra, is much lower than that in the surrounding photosphere, allowing us to see about 1,000 km (600 miles) further into it. However, sunspots do not always show the circular outline shown above, with the penumbra concentrically surrounding the umbra. Often the penumbra is more elongated on one side than on the other, and this can itself cause a "Wilson Effect" when the spot is near the limb. For example, if the penumbra shows more elongation towards the west, this will remain visible as the sunspot approaches the western limb, giving the impression that the spot is a depression. Conversely, a spot with a penumbra displaced towards the east can sometimes show the opposite of the Wilson Effect: penumbra broader towards the centre of the disc, and *narrower* towards the limb. For many years astronomers thought that spots with this appearance were small mounds above the solar surface. The Wilson Effect is generally only visible in symmetrical sunspots and relatively simple sunspot groups and is much harder to see in more complex groups.

A sunspot begins life as a tiny dot about 3″ or less in diameter known as a *pore*. Pores are not true sunspots and have a lighter and greyer appearance than genuine sunspot umbrae; as we shall see in Chapter 5, they are not included in counts of sunspot activity. A pore becomes a sunspot proper when it becomes as dark as a sunspot umbra and grows somewhat larger. Such a spot has no penumbra and borders immediately on granulation on all sides. Small spots such as these are the most common type of sunspots. A spot in its early stages of development is often surrounded by faculae, mentioned above when we discussed limb darkening. Indeed, faculae often herald the formation of sunspots. If you see a prominent clump of faculae at the eastern limb, it would not be surprising to see a small spot in the same region within the next day or two.

The smallest spots often disappear after a day or less, but if the spot survives then over the next two or three days the initial spot is joined by another spot close by, the two forming a "bipolar" sunspot group – that is, a group of two spots of opposite magnetic polarity. The preceding, or westernmost spot is known as the "leader" and the trailing spot the "follower". The separation between the two spots gradually increases. The leader and sometimes the follower may go on to form penumbrae and by the end of a week a number of smaller spots may have erupted between the two main spots, the group as a whole perhaps containing ten or more individual small umbrae. If the group's magnetic field is complex, as often happens near solar maximum, the whole complex may grow even larger and many more small spots may erupt within it. The very largest groups can contain over 100 spots. Regardless of the size of the group, the larger umbrae, including the leader and follower, often split into two or more sections. The splitting of a large spot is often heralded by a deep notch appearing on one side of the umbra's longer dimension. Within a day or less, this notch has cut across the spot and split it into two sections. The brighter channel between the two sections is known as a *light bridge* and is often as bright as the photosphere. An occasional variant on the light bridge is a bright "island" of photosphere in the middle of an umbra. If an umbra suddenly splits up into several sections this suggests considerable activity and the possibility of flares emanating from the group (see later in this chapter and also Chapter 6).

Most sunspot groups reach their maximum size and complexity when they are between a week and two weeks old. After that time even a large group starts to decay. The umbrae of the spots get smaller and more and more of the group's area is covered with penumbrae. If the group is near the limb when at this stage, it will be surrounded by extensive faculae. However, the leader spot tends to remain prominent, with a well-defined umbra and penumbra, even when the rest of the group is dying out.

The longevity of sunspots and groups is as variable as their sizes. The smallest spots and pores last for only a few hours or days, while very large groups can last for weeks or even months. Because a sunspot takes only about two weeks to cross the Sun's disc, it follows that a long-lived spot can reappear at the eastern limb a fortnight after it has disappeared in the west. This is often the case with a large group of the type mentioned above. When it reappears in the east it has often decayed considerably, the leader and perhaps the follower still being present but only faculae and penumbrae visible between them. Such groups have been known to survive several rotations of the Sun, getting smaller each time they come round. A good example of a large, long-lived

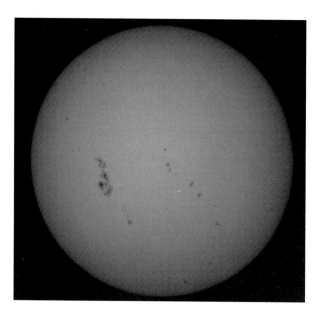

Figure 3.4. The giant sunspot group of March 2001, McIntosh classification Fkc. This was one of the largest sunspot groups ever recorded and was associated with widespread auroral displays. Photograph by the author, using an 80 mm refractor, Baader AstroSolar Safety Film aperture filter and eyepiece projection giving an effective f/ratio of f/33 (see Chapter 7). Exposure 1/250 second on Kodak Elite Chrome 200 film.

group occurred in the spring of 2001 (Figure 3.4). At the end of March the group was a spectacular complex, dominating the field of view at high magnification. It produced some powerful flares, and brilliant displays of the aurora were reported from many parts of the world. The group came round again in April and once more in May, but its size and number of spots were much reduced. Even very large groups are not always long-lived, however. Ten years earlier, for example, again in March, there was a similar huge group which produced flares and aurorae, but it did not reappear after it went round the limb of the Sun.

Sunspots are classified into types according to the stage they have reached in their evolution. Astronomers classify sunspots using a three-letter code. This so-called "McIntosh scheme" is a modification of the "Zurich" scheme originally devised at the Swiss Federal Observatory in the mid-twentieth century and is now the standard scheme throughout the world for classifying the visual appearance of sunspots. It is useful to know how the scheme works, as it is often used to describe spots and groups in reports and it makes it easier to identify the different types of spots as they appear on the Sun.

The first letter of the code is written in upper case and describes the basic type or class of the spot group:

A   Single spot or very small group, with no penumbra
B   Bipolar group, with no penumbrae
C   Bipolar group. One spot (usually the leader) has a penumbra
D   Bipolar group with penumbra around both leader and follower
E   Group between 10° and 15° long with penumbrae around both leader and follower and many small spots between leader and follower
F   Group more than 15° long with penumbrae around both leader and follower
H   Single spot with penumbra

The second and third letters are both written in lower case. The second describes the penumbra of the group's largest spot:

x   No penumbra
r   Penumbra partly surrounds largest spot
s   Small, symmetrical penumbra, north–south extent less than $2\frac{1}{2}°$

a   Small, non-symmetrical penumbra, north–south extent less than $2\frac{1}{2}^{\circ}$

h   Large, symmetrical penumbra, north–south extent more than $2\frac{1}{2}^{\circ}$

k   Large, non-symmetrical penumbra, north–south extent more than $2\frac{1}{2}^{\circ}$

The third letter describes how tightly grouped the spots are in the central part of the group, between the leader and follower:

x   No spots between leader and follower. Applies only to types A and H, where there is only one spot in the group

o   Few or no spots between leader and follower

i   Numerous spots between leader and follower, none of them with fully fledged penumbra

c   Many large spots between leader and follower, at least one of which has a mature penumbra

The first letter describes the stage the group has reached in its development. As we have already seen, when a group starts life by turning from a pore into a sunspot proper it is a single spot without a penumbra – i.e. class A. Over the next few days it becomes a small bipolar group without any penumbrae (class B). If the leader and then the follower develop penumbrae it becomes class C and then D respectively. If the group grows and becomes more complex after this point it may become an E or even an F group. As the group decays and leaves behind only the leader spot, it becomes type H (Figure 3.5).

Let us see how the full three-letter description works in practice. The most common type of group is also the simplest – type Axx, a single spot with no penumbra and, by definition, no spots between the leader and follower. A more complex example is type Dki – a bipolar group with penumbrae around leader and follower, the main spot having a large, asymmetric penumbra and numerous small spots (without penumbrae) between the two main spots. One of the most spectacular types is Fkc – bipolar group with large, asymmetric penumbra around its main spot and with many substantial spots between the two main spots.

Note that although the McIntosh scheme covers a great variety of sunspot types, not all combinations of letters are possible. For example, if we consider the descriptions above, we can see that, say, types Fxx or Aki would be logically, as well as physically, impossible!

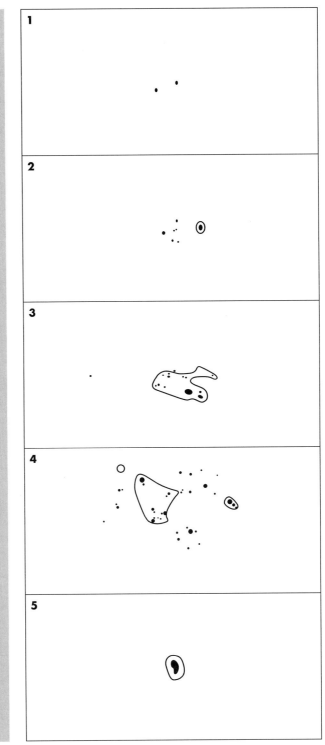

Figure 3.5. Some examples of McIntosh sunspot group classifications. (1) Bxo; (2) Csi; (3) Dki; (4) Ekc; (5) Hsx.

Faculae

Faculae are areas of gas about $300°K$ hotter than the photosphere, forming a little higher up than sunspots. Nearly all sunspots have at least some faculae associated with them, but the faculae are usually impossible to see when they are near the centre of the solar disc and only become distinct when near the limb, where they show up by contrast with the limb darkening.

You may notice that some areas of faculae contain no sunspots and appear to exist in complete isolation from any spots or groups. Actually, however, faculae are nearly always associated with sunspots. As mentioned above when we discussed the formation of sunspots, an area of faculae often forms before a sunspot or group of sunspots appears in the same position. Similarly, faculae may persist for many days or weeks after a group has decayed and disappeared. With the notable exception of one type of faculae, all faculae form in the same zones of solar latitude as sunspots – i.e. two bands parallel to the equator, stretching to $40°$ or so. In fact, astronomers believe that faculae are regions produced by magnetic fields too weak to cause sunspots.

Like sunspots, faculae come in a large variety of shapes and sizes. They can range from small, bright "blobs" to large, extended patches, while some can appear as long, thin streaks. Some extensive areas of faculae, as are often found around active spot groups, have exceedingly complex arrangements. Their low contrast makes them difficult to observe, but one property can make them stand out better. Faculae emit very strongly towards the blue end of the visible spectrum. As noted in Chapter 2, Mylar-type filters transmit strongly in this part of the spectrum. This gives the Sun a blue appearance as seen through the eyepiece but many observers have noted that faculae appear more distinct through these filters than with glass filters.

Polar Faculae

The exception mentioned above to the fact that faculae form in the sunspot latitude zones is the *polar faculae* – faculae that form at solar latitudes higher than about

50° and most frequently between 75° and 85°. They are not nearly as easy to see as normal, spot-related faculae. Unlike the latter they do not occur as extended patches but as tiny, isolated bright spots a few arc seconds across, often about the same size as individual pieces of granulation, therefore requiring good seeing to be visible. They are also very short-lived, never lasting for more than a few days and often with lifetimes measured in minutes. Adding to the difficulty of observing them is the fact that, as we shall see in the next chapter, the position angle of the Sun's poles varies, so that at some times of the year one pole is tipped towards us, while at other times the other pole is favourably positioned. It follows that faculae are most easily observed around the north solar pole at some times and around the south pole at others.

Polar facula activity varies in reverse to the activity of sunspots and their associated faculae, in that polar faculae occur in greatest numbers during solar minimum and the years immediately preceding it. During the rise back to maximum they disappear and they are relatively rare at maximum.

Polar faculae have not been so thoroughly studied as other features of the Sun and systematic investigation of them is useful work for the amateur. Although they are much harder to see and keep track of than sunspots and sunspot-related faculae, I have often seen them with just a 60 mm refractor and the projection method with a well-darkened projection box.

Flares

A solar flare is by far the rarest and most exotic solar phenomenon you can observe with your telescope in ordinary visible light. As I mentioned in Chapter 1, flares are extremely powerful in radio and X-ray wavelengths, where they can sometimes outshine the entire Sun for short periods, but they emit relatively little energy in the visible spectrum. Flares can be detected in certain narrowly defined wavelengths of visible light, using special filters which isolate these wavelengths from the rest of the spectrum (see Chapter 6), but in everyday white or "integrated" light all but the most intense flares are invisible. When a flare is visible in white light, it is known as a *white-light flare*, and if you see one it is a major event that needs to be reported. More details on reporting are given in Chapter 5.

In fact, the first solar flare to be discovered was a white-light flare. It was observed on the morning of 1859 September 1st, long before instruments for observing the Sun in radio, X-rays or narrow-band visible wavelengths had been invented. Two British amateurs, Richard Carrington and Richard Hodgson, were observing the Sun independently of each other when they noticed two brilliant patches of light suddenly appear in the midst of a large sunspot group. The phenomenon faded away and disappeared from view in a matter of five minutes. Shortly afterwards, early magnetic instruments recorded a serious disturbance in the Earth's magnetic field and there was also a brilliant display of the aurora. Some astronomers immediately inferred the connection between solar activity and magnetic storms on Earth, although it was not until near the end of the nineteenth century that the solar origin of aurorae was widely accepted.

White-light flares are extremely rare. Only a handful have been recorded since the original Carrington–Hodgson observation of 1859 and I, personally, have never yet seen one in more than a decade of regular solar observation. Seeing one demands a great deal of luck, but some characteristics of white-light flares can improve your chances of seeing them. First, flares usually occur only in large and complex sunspot groups – i.e. classes D, E and F in the McIntosh system mentioned above. They are especially likely to occur in groups with a compact structure and containing many spots. Sub-classes ki (largest spot has large, asymmetric penumbra and numerous spots between leader and follower) and kc (same, but spots between leader and follower have penumbrae) are the most flare-productive. Another way of increasing your chances of seeing one is to monitor an especially complex group several times a day. Your chances are much reduced if you observe just once a day, as this provides just a snapshot of the group's behaviour. Also, like faculae, white-light flares emit much of their visible light towards the blue end of the spectrum, so using a blue-transmitting Mylar filter can help as well.

White-light flares show up as small, bright spots, patches or streaks within a sunspot group. However, be careful not to confuse a light bridge with a flare. White-light flares usually have lifetimes of less than ten minutes before fading away, whereas light bridges retain the same brightness for hours or even days. The bright areas of photosphere that occasionally appear in the umbrae and penumbrae of sunspots can also be confused with flares.

Chapter 4

Solar Drawings and Position Measurements

The cheapest and most convenient way of recording the Sun's appearance each day is to draw it. Drawings can also be important in determining sunspot statistics, as I shall describe in Chapter 5. Photography and CCD imaging are in theory more accurate than drawing, but they are less practical, for several reasons. It is not as easy to define the cardinal points (north, south, east and west) on a photographic image as accurately as on a good drawing. Secondly, it is difficult to derive precise positions of sunspots from a photograph. Thirdly, a photograph taken at a scale small enough to show the whole Sun on one frame – essential for measuring the positions of sunspots – has a relatively limited resolution and small spots which can be picked up visually can be missed on a photograph. Photographs taken at a small scale such as this sometimes suffer from distortion towards the edge of the frame, which can compromise the accuracy of any sunspot position measurements. Finally, a drawing is ready for analysis as soon as it has been completed, whereas photographs and electronic images both have to be processed. This can be time-consuming, especially if you are sending results in to an amateur observing organisation that requires observations on a monthly basis.

Drawing Using the Projection Method

I do my solar drawings using a method employed by amateur solar observers for many years. In my projection box I insert a card containing a finely ruled grid. Underneath the paper on which I make my drawing I clip an identical grid, whose lines are thick enough to show through the drawing paper. I then simply copy the positions of any sunspots present from the grid in the projection box onto the grid under my drawing paper, and I have an accurate plot of what was visible on the Sun that day. This potential for accurate drawings is a major reason why I prefer the projection method to observing by direct vision through a filter. Using the latter method, the result is bound to be less accurate, as the drawing has to be done freehand. You could compromise by using a crosshair eyepiece, such as that used on a finderscope, and so divide the Sun's disc into quadrants. It may also be possible to use a reticule intended for a microscope or a precision measuring instrument, obtained from an optical surplus supplier. But in any case, I still consider the projection method the most convenient, as it allows you to comfortably look at the Sun's image with both eyes and involves no eye strain as you switch your eyes back and forth from the image to the drawing.

Making a Projection Grid

Figure 4.1 shows an example of a typical solar drawing grid, consisting of a circle to fit the Sun's projected image and divided into small squares. These squares are further divided by diagonals for more precise plotting. Many observers use a circle 152 mm (6 inches) in diameter for their solar drawings, as this is a standard solar disc size used by many organisations when measuring solar drawings. If you have a very small telescope, however, and therefore use a smaller projection box, you may wish to use a 100 mm (4 inch) disc. A 6 inch disc is best divided using Imperial measurements into half-inch squares. If you prefer to use a 4 inch disc, 4 inches is actually 102 mm, but for accurate plotting you need somewhat smaller squares, and so it is best to use metric units. Drawing the disc to

Figure 4.1. A solar
drawing grid, consisting
of a 152 mm (6 inch)
circle divided into $\frac{1}{2}$ inch
squares, as used in the
projection box and
underneath the drawing.
The 5° intervals marked
out around the
circumference are for
determining the Sun's
true orientation for
sunspot counting
purposes (see
Chapter 5).

a diameter of exactly 100 mm and dividing it into
10 mm squares gives good results.

Constructing your own grid is quite easy, although it
takes care and patience, since it has to be accurate. In
fact, you need to make two grids: one for the telescope
and one for the drawing. To start, you need a ruler and a
360° protractor for the size of disc which you wish to use
(good stationery shops should supply both 4 inch and
6 inch protractors). You also need a sheet of Bristol
Board – the same material as that used for making the
screen in a projection box (see Chapter 2). On the Bristol
Board draw a circle of the appropriate diameter. Then,
using the protractor, mark four points 90° apart around
the circumference of the circle. These will later be the
north, south, east and west points when we have worked
out which way round the telescope shows the solar
image. Join these points to form a cross over the circle.
Its centre should be at the exact centre of the circle.

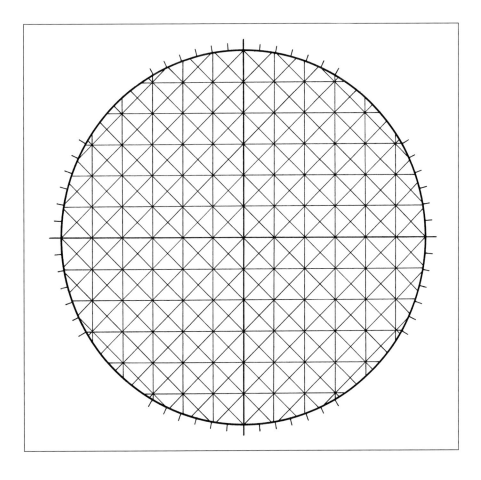

Then, mark a series of points on the horizontal axis of the cross, half an inch apart for the 6 inch circle, 10 mm apart for the 4 inch. Do the same with the vertical axis. Now, move the ruler 30 mm or so above the horizontal axis and again mark a series of points half an inch apart, so that their positions precisely correspond with those on the axis. Do this for each axis. For even greater accuracy, you could repeat this procedure 30 mm below each axis. Finally, for each axis, join up the corresponding points with lines stretching across the circle, taking time and care to ensure the lines are precisely parallel. When you have finished, you should have a grid composed of squares of the appropriate size. It is then quite easy to draw in the diagonals, using the corners of the squares as guides. Some observers like to letter their squares – for example, alphabetically on the horizontal axis and numerically on the vertical. This helps you remember on which square you need to plot a sunspot. Otherwise, you can simply count the spot's location in terms of the number of horizontal and vertical squares from the centre of the disc.

When drawing the grids, you should do everything in pencil first of all, so that you can correct any mistakes. The grid to go in the projection box should be drawn with very faint pencil lines, since heavy lines are a distraction and could cause you to miss some of the smaller sunspots. However, the grid to be used under the drawing needs to have black ink lines, so that it shows up clearly through the drawing paper. To ensure the best possible accuracy, the lines should not be too thick. A fine-tipped drawing pen should produce lines thick enough to be visible through ordinary paper. The type of paper you use is a matter of personal preference, but the thinner the paper the better, since this allows the lines to show through more clearly. "Bank" quality paper, as used for typing carbon copies, is excellent, but is becoming less common now in the age of the word processor. A good alternative is tracing paper, although this is more expensive per sheet than bank paper. Some observers use a notebook for their drawings, with the grid paperclipped behind one of the sheets, while others do their drawings on loose sheets of paper on a clipboard. Whichever method you choose, make sure that both the grid and the drawing paper are steadily mounted and cannot move relative to each other. If you use a clipboard, use a clip at one side of the paper in addition to the main clip at the top of the board.

In order to orient the image correctly you need to be able to rotate the projection grid. You can do this by rotating the entire box, but I find it easier to rotate just the grid. To make the grid rotatable, cut it out to a circular shape and attach it to the screen end of the box with a pin or small screw through the central axis of the grid.

When you have placed the pencilled grid in the projection box, adjust the box so that a focused image of the Sun exactly fits the 6 inch circle. If you are observing near sunrise or sunset the Sun's image may appear slightly elliptical, owing to atmospheric distortion, and so these are not good times to make accurate drawings. The projection screen with its grid needs to be precisely square-on with respect to the telescope's "optical axis" – i.e. the path a beam of light takes running through the dead centre of the tube. If you look carefully at the projected image, preferably with the grid temporarily removed and a plain white screen inserted in its place, you should see at least parts of the edge of the field of view near the corners of the screen. If the edge of the field is concentric with the centre of the projection grid and is central within the box (i.e. is at an equal distance from all four corners of the screen), then the box is on-axis. If the screen is notably tilted away from the axis, the Sun's image is distorted into an ellipse and so the accuracy of any sunspot positions obtained is compromised. You need to look out for this if your projection box is mounted on its own stand separately from the telescope, for then the telescope is moving all the time with respect to the box.

Orienting the Image

Before you can begin drawing, you need to establish the four points of the compass on the image. Establishing the east–west direction is a simple matter. Using the telescope's slow-motion controls, place a sunspot on the horizontal line running through the centre of the grid. Switch off the telescope's motor drive (if it has one) and watch the direction in which the spot drifts across the screen. Then rotate the projection grid so that the horizontal line runs along this direction of drift. Repeat this process until the spot drifts precisely along this line. The grid is then orientated east–west. The Sun, like everything else in the sky, moves from east to west, so the mark on the circumference of the circle towards which the spot is moving is the west point. You can mark this point "west" for future reference.

You then need to establish which end of the grid's vertical axis is north. If you have an equatorial mount you can do this by turning the declination slow motion so that the telescope moves northwards, towards where Polaris would be if it was visible. If you look at the projection screen, the direction in which the view appears to be moving is north, and you can mark the north point accordingly. If your telescope is an alt-azimuth, this exercise is best performed near midday, when north is vertically above the Sun in the sky. Whatever type of telescope you use, you will probably notice that north is at the top of the image, in contradiction to the rule that an astronomical telescope gives an upside-down image. This happens because projection involves an extra reflection, which "flips" the image north–south. The effect is visible in a refractor or a reflector. East and west, however, are not affected, and remain the same as in an astronomical telescope – i.e. west to the left, east to the right. However, if you are using a refractor and are projecting the image through a star diagonal, as I employ in my projection setup, east and west *are* flipped, putting east to the left and west to the right, exactly the same as the naked eye view of the Sun. If you are using a Schmidt–Cassegrain or Maksutov telescope, you will be viewing a filtered solar image directly through a star diagonal. In this case the image orientation is the same as that of a projected image without a star diagonal – i.e. west left, east right. The diagram in Figure 4.2 shows the orientation of the Sun's image as seen in different telescope setups.

Making the Drawing

You are now ready to begin the drawing. Plot the sunspots using an HB or H grade pencil. Too soft a grade of pencil will quickly wear down and become too blunt to give an accurate rendition, whereas too hard a grade makes it difficult to rub out mistakes (as you will probably want to do quite frequently if you are just beginning!). I like to start with the western part of the northern half of the Sun's image and then work eastwards, before switching to the southern half and working westwards. Don't get bogged down in trying to reproduce the finest details in your drawing. The most important aim in a full-disc drawing is to get the

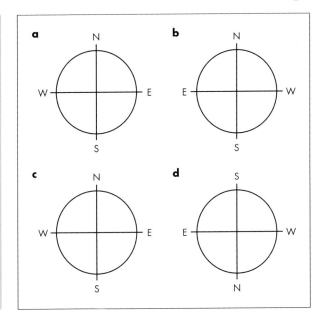

Figure 4.2. Orientation of the Sun's image as seen in various telescope setups. **a** Straight projection through a refractor or Newtonian. **b** Projection using a star diagonal. **c** Observed directly in a SCT or Maksutov telescope. **d** Observed through a coronagraph (through a star diagonal, from above – see Chapter 6).

positions of the spots correct. Draw in enough detail to show the general shapes of the spots and groups, with the umbrae and penumbrae of the principal spots. If you are drawing a large group, start by establishing the positions of its principal spots on the grid squares and diagonals, and then fill in the finer details between them, using the main spots as a guide.

If your telescope is on an alt-azimuth mounting, remember that the Sun's image is slowly rotating, because the mount is not aligned with the Earth's axis. Therefore you need to adjust the orientation every few minutes to keep the east–west line of the grid coincident with the east–west direction on the Sun's image. This is not a problem with an equatorial. Provided that the mount is aligned with the pole reasonably accurately, the orientation should remain the same throughout the observation. Best of all is an equatorial mount with a motor drive, as then you do not have to keep adjusting the RA slow motion control to keep the Sun on the circle. However, even with a well-aligned equatorial and a good drive, the Sun can still stray slightly off the grid, due to small errors in polar alignment and the drive system. You still, therefore, need to make occasional adjustments using the slow-motion push buttons on the drive control, but these are much smaller and less frequent than with an alt-azimuth or an undriven equatorial. It is also worth checking the orientation of the image at the end of the observation as well as at the beginning, just in case you have accidentally jogged the projection screen, as sometimes happens.

When you have drawn all you can see on the full-disc solar image, place a plain piece of Bristol Board or other white paper in the projection box over the grid and insert a higher-power eyepiece into your telescope. I use a 7 mm Orthoscopic on my 80 mm refractor, giving a magnification of 130×. Now carefully scan the enlarged solar image and look for any small spots you may have missed on the smaller image. When you see a hitherto unrecorded spot, make a mental note of its position relative to other sunspots, switch back to lower power and record its position on the full-disc drawing. More often than not, such missed sunspots are large enough to be easily visible on the smaller image once you know where to look for them. Sometimes a spot is big enough to make you wonder how you missed it in the first place! When scanning at high power, pay particular attention to the east and west limbs of the Sun, where it is easy to miss small spots coming onto or moving off the Earthward side of the Sun. Often an area of faculae seen near the limb at lower power turns out to have one or more spots embedded within it when examined with higher magnification. Also deserving close scrutiny are the areas between the main sunspot latitude bands and the poles, where high-latitude sunspots occasionally appear, especially near the minimum of the sunspot cycle.

When you have finished scanning the image at high power and recorded any missed spots, the drawing is complete. The final stage is to record some essential details: the date on which you made the drawing, the time at which you completed it, the observing conditions and the instrument used. Astronomers write the date in successively smaller units, e.g. 2002 October 4th. The time should always be in Universal Time (UT), which is the same as Greenwich Mean Time. If Summer Time is in use or if you live in a different time zone from GMT, remember to subtract the appropriate number of hours from the time shown on your watch. You can record the conditions in words, e.g. "clear, steady", "boiling", "passing clouds", etc. Some observers grade the quality of seeing on a scale of 1 to 5, similar to the Antoniadi seeing scale used by planetary observers. In this scale, 1 is perfect or near-perfect seeing with hardly a ripple in the image (very rare from most sites), while 5 is extremely poor seeing, with the image "boiling" violently and making a drawing hardly worthwhile. Some solar observing organisations also ask for observers to include the current *rotation number* with their drawings. This is a system of marking the Sun's rotation and was begun in the nineteenth century by Richard Carrington (who was also co-discoverer of the first solar flare, as I mentioned in Chapter 3). Carrington's system uses a rotation period of 25.38 days, with rotation number 1 beginning in 1853. Now, at the beginning of the twenty-first century, the Sun is approaching the 2000th rotation number since the commencement of Carrington's system. The main use of the Carrington system is to determine the heliographic (i.e. solar) longitude of sunspots, as I shall explain shortly.

Deriving Sunspot Positions

A completed solar drawing is useful not only for illustrating what was present on the Sun that day, but also for deriving the heliographic positions of sunspots and groups. From the latter we can learn certain characteristics of the sunspot cycle and of the sunspots themselves.

Sunspot positions can be derived from drawings in two ways – by calculation or using specially designed grids. The mathematical method can produce accurate results, but it is very laborious. The second method, which I shall describe here, is much easier. The most common type of measuring grid is known as the Stonyhurst Disc. A Stonyhurst Disc consists of a 6 inch circle with the cardinal points marked, as on our drawing grid, but instead of horizontal and vertical lines it is ruled with lines of latitude and longitude. The vertical lines (longitude) become more curved with distance from the centre of the disc, reminding us that the Sun is a three-dimensional object. To measure sunspot positions, a completed solar drawing is laid over the Stonyhurst Disc and the latitudes and longitudes of the spots are simply read off.

Unfortunately for the solar observer, position measurements are not as simple as this. The orientation of the solar image that we so carefully established above when making the drawing is only the orientation *as seen from Earth*. The Earth's axis is not perpendicular to the Earth's orbit but is tilted 23.5° to the perpendicular. This, of course, is the cause of the seasons, but it also causes the position of the Sun's poles to vary. Similarly, the Sun's own axis of rotation is inclined at 7.25° to the perpendicular to the Earth's orbit. As seen from Earth, the axial tilts of the two bodies combine to produce two effects. First, during the year the position of the Sun's north pole varies by 26.3° either side of the apparent north point of the disc. Secondly, the Sun appears to "nod" towards and away from us by 7.2°. We must correct for both if we are to establish accurate sunspot positions.

Fortunately, both effects are regular and can be predicted accurately in advance. The changing angle of the north solar pole to the apparent north point of the solar disc is known as the *position angle* and is usually abbreviated to P. When the north pole is on the east side of the apparent north point it has a positive value and when on the west side it is negative. In early January the Sun's axis coincides with the apparent north–south line and so the value of P is 0°. The axis then gradually moves west of the apparent north point, reaching its maximum westward tilt of 26.3° (i.e. –26.3°) in early April. After April the axis backtracks, returning to the 0° value in early July. It then tilts increasingly eastwards, reaching its maximum eastward value of +26.3° around 11th October, before moving westwards again and returning to the zero point in January. Put more simply, the value of P is negative (i.e. west) in the first half of the calendar year and positive (east) in the second half.

The extent of the Sun's "nodding" towards or away from us is defined as the heliographic latitude of the apparent centre of the disc and is abbreviated B_0 (pronounced B-nought). The value of B_0 does not vary in synchronisation with P. In early December the solar equator coincides with the centre of the disc and so B_0 is 0°. In the early months of the new year the south solar pole tilts towards us, giving B_0 a negative value. The south pole reaches its maximum tilt towards us of 7.2° in March and then begins to tilt away from us, passing through the zero point again in early June. For most of the latter half of the year the north pole is tilted towards us and B_0 has a positive value, reaching its maximum of +7.2° in September. Thus at most times of the year the lines of solar latitude – and the apparent paths followed by sunspots as they rotate across the disc – are curved, because the poles are at an angle to us. Only during the brief few days in June and December do sunspots follow straight paths across the disc. Therefore solar observers use a set of eight discs, one for each degree of B_0. Note also that P and B_0 have one feature in common: the rate of change of the value of both angles is slowest when they are at their maximum and fastest when they pass through the zero point. For example, P is more than $-26°$ for over three weeks from late March until mid-April, whereas in January it varies by nearly $5°$ (passing through $0°$ on January 5th) during the first ten days of the month.

A knowledge of P and B_0 at the date of observation is sufficient for calculating the latitudes of any spots present. However, to calculate the longitudes of spots we need a third statistic. Solar astronomers measure longitudes using the longitude of the *central meridian*. This is simply a line drawn across the disc from the position of the north end of the rotation axis (as determined by the value of P) through the centre of the disc and is the point at which sunspots and other solar features pass across, or "transit" the disc's centre. The longitude of the central meridian (abbreviated L_0) is $0°$ at the beginning of each new solar rotation, as determined by Carrington's system described above. Heliographic longitude is measured from east to west, but the Sun also rotates from east to west, and so the longitude decreases with time, progressing from $0°$ to $350°$, then to $340°$, and so on.

To demonstrate how to work out P, B_0 and L_0, let us use one of my own drawings, made at 11:30 UT on 1996 November 23rd (Figure 4.3), as an example. I have deliberately chosen a drawing made at sunspot minimum, in order to keep things simple. At maximum, when there are often many large, complex groups on the disc, measuring the positions of all the sunspots would take much longer.

To begin with, we need to find the three essential statistics for the date and time at which the drawing was made: P, B_0 and L_0. These are tabulated in several annual publications, including the *Handbook of the British Astronomical Association* and the *Astronomical Almanac*. Both publications are released annually. The *Astronomical Almanac* lists these figures for each day of the year, while the BAA *Handbook* gives them at five-day intervals, plotted for the previous midnight, i.e. 0 hours UT. P and B_0 change slowly enough that we can use just the daily figure, and so users of the *Almanac*

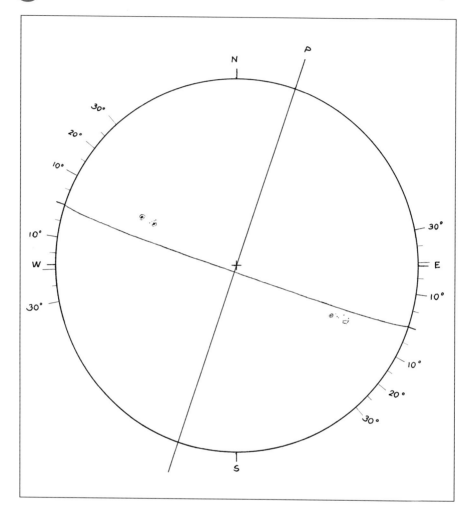

Figure 4.3. Whole-disc drawing, made by the author on 1996 November 23rd, showing two sunspot groups. The position angle *P* is marked, as well as the true location of the Sun's equator and latitudes.

can simply copy down the values for the date of the drawing. If you use the BAA *Handbook* you need to interpolate between the two nearest values. Because the BAA *Handbook* is the commonest reference source in the UK, where I live, I will use this in our example.

As it happens, 1996 November 23rd, the date of our example drawing, falls on one of the BAA *Handbook* five-day intervals. The *Handbook* says that *P* is +18.9° and B_0 is +1.8°. Be very careful to get the signs right, as otherwise the results would be hopelessly wrong – although, as discussed above, you can generally reckon on *P* and B_0 being negative in the first half of the year and positive for most of the second half.

Because the Sun rotates through a full 360° in 25.38 days L_0, the longitude of the central meridian, changes quickly enough that its value for the time of day on which you made your observation is significantly different from the midnight value. L_0 decreases by 13.2° per day, which means that it decreases by 0.55° per hour. The "Sun" section of the *Handbook* includes a table headed "Decrease of L_0 with time". From this we can see that during the first ten hours of the day the value has dropped by 5.5°. To get the amount of decrease for 11:30, we need the appropriate value for 1.5 hours, which from the table is 0.8°. Therefore the total decrease in L_0 since midnight is 6.3°. Subtracting this from the midnight value gives us 208.3°. Because we need to refer to them, it is useful to make a note of these values on the sheet on which the drawing is made.

The next job is to mark the Sun's axis on the drawing. We have found the value of P to be +18.9° – that is, the north end of the Sun's axis is inclined 18.9° east of the apparent north point of the disc. Mark the appropriate angle on the drawing using the 6 inch protractor and draw an accurate line across the solar disc through this mark and the cross at the centre of the disc. In the example given here this has already been done. A 6 inch protractor will not allow you to measure to one-tenth of a degree, so just mark the angle to the nearest half-degree. We are now ready to use the Stonyhurst Discs. Select the disc whose B_0 value is the closest to the B_0 value for the day of your observation. In this case, $B_0 = 1.8°$, so we should choose the disc for 2°. Lay your drawing over the disc, turning it round so that the Sun's axis which you marked on the drawing coincides with the central, straight north–south line on the disc. Make sure that your drawing disc coincides exactly with the edges of the Stonyhurst Disc, and then clamp the two sheets firmly together with a paperclip at each side.

It is very satisfying to look at your drawing at this point, because the Stonyhurst Disc's curved lines of latitude and longitude give it a three-dimensional appearance, reminding us that we are drawing and measuring positions on a sphere. Because the line of the equator is accurately established, we can tell instantly which hemisphere a spot is located in. On this particular day two reasonable-sized sunspot groups were visible, one in the northern hemisphere and the other in the south – proof that even at solar minimum there are often interesting things to see on the Sun. Reading off the latitudes of the spots is now a simple matter, because the latitudes are marked off at 10° intervals on the disc and for every degree at the limb.

Measure the latitude of a spot using the centre of its umbra. You can use a ruler graduated in millimetres to measure the latitudes more accurately. For example, the leader sunspot of the group in the northern hemisphere is 8 mm above the line representing the equator. On the latitude scale at the limb this translates into 6°, and so this spot has a latitude of 6° N, or +6°. Measuring in the same way, we find that the follower spot of the same group has a latitude of +5°. In the southern hemisphere group, both the principal spots have latitudes of –3°.

To measure the longitude of a spot we need to know its distance from the central meridian – i.e. the north–south axis line. On the Stonyhurst Discs the longitudes are marked off at intervals of 10°, as for latitude. There is no scale measuring longitude to the nearest degree, but it is usually fairly easy to interpolate longitudes using a millimetre ruler. The essential thing to remember when measuring longitudes is that they run from *east to west*, so longitudes of spots west of the central meridian are higher than that of the central meridian (i.e. L_0), whereas those for spots east of this line are lower. Thus in our example drawing, to find the longitude of the leader spot of the northern hemisphere group, which is west of the central meridian, we need to measure its distance from the line (36°) and add this value to L_0 (208°), giving a heliographic longitude of 244°. To find the longitude of the equivalent spot in the southern hemisphere group, we have to *subtract* its distance from the line (again 36°) from 208°, giving a longitude of 172°.

The Stonyhurst Disc method is much the easiest way of measuring sunspot positions and can, in theory at least, give results accurate to 1° for spots near the centre of the disc. Results are less accurate, however, for spots close to the limb. You can obtain a set of Stonyhurst Discs from the solar observing section of a national astronomical society and there is at least one Web site from which you can print them out. One alternative method which you may read about in astronomical literature is known as "Porter's Disc". This is a 6-inch disc divided into small squares, from which the horizontal and vertical positions of the spots are measured. These *x–y* coordinates are then translated into heliographic longitude and latitude using trigonometry. I do not recommend this method, as it involves a considerable amount of calculation and is much less convenient than the Stonyhurst Disc method. In fact, there are nowadays some computer programs which will work out the precise values of P, B_0 and L_0 for the date and time of your drawing and even the heliographic coordinates of the spots if you enter their x–y positions (in millimetres) measured from your drawings. However, the latter still have to be measured precisely and you may find it just as easy to get good positions from the Stonyhurst Discs.

What We Can Learn from Drawings

We noted in Chapter 3 that sunspots form in distinct "bands" of latitude on either side of the equator – generally between 5° and 30°. Spots at latitudes of higher than 40° are quite uncommon and they never form at or near the poles. But these bands are not constant. If we make drawings of sunspot positions over several years and plot the latitudes of the spots on a graph, we find that the average latitude at which spots form gradually decreases as the solar cycle progresses. Early on in a new solar cycle, shortly after minimum, spots tend to appear at latitudes well above 20° and often at 30° or even higher. By the time of sunspot maximum the average latitude of sunspots has declined to about 15° and at minimum the small number of spots seen at this stage of the solar cycle form within a few degrees of the equator. However, by the time activity "bottoms out" at minimum, occasional spots are seen at high latitudes again, heralding the start of the new solar cycle. In this respect there is always a certain amount of overlap between one cycle and the next. This phenomenon of gradual decline in sunspot latitudes was discovered in the nineteenth century by (yet again) Richard Carrington. Further studies into it were undertaken by Gustav Spörer in Germany and today it is known as *Spörer's Law*.

The drawing in Figure 4.3 used above to describe how to measure sunspot positions provides a good demonstration of Spörer's Law. I made this drawing around the time of minimum activity in late 1996 and both the sunspot groups present are at low latitudes: the northern hemisphere one at +6°, the southern one at –3°. This is exactly what we would expect at this stage of the solar cycle. I did not see any high-latitude spots of the new cycle on that day, although I had noted the first one of these as early as June of that year. Solar minimum can be a trying time for the solar observer, since there are sometimes no sunspots for days or weeks on end and, with the exception of the polar faculae, there is often very little to look at on the Sun. It is therefore rather exciting to see and record the first high-latitude sunspots, as it tells us that the next solar cycle, and therefore greater activity, is just around the corner.

For convenience, astronomers refer to individual sunspot cycles using a sequential series of numbers, starting at solar cycle 1, which reached maximum in 1761. Choosing this cycle as number 1 was an arbitrary decision by astronomers, as of course activity was varying in cycles before then. Now (2002) we are at the maximum of solar cycle 23. The current cycle began in 1996 (i.e. at the last minimum) and is expected to end when activity next bottoms out, probably around 2007, if the cycle turns out to have an average length – i.e. 11 years. Only time and patient recording will tell.

Measurements of the longitudes of sunspots can also be useful (Figure 4.4). For example, as we saw in Chapter 3, some large and active sunspot groups survive more than one passage across the Sun's visible hemisphere. If we know such a group's latitude and longitude on its first rotation, we can very quickly establish whether or not it is the same group when it reappears at the eastern limb on its second passage. It is very rewarding to prove using your own drawings that a group is visible the second time round. Sometimes the second appearance of a group is obvious from its general shape and approximate position on the Sun's limb, plus the knowledge that it disappeared round the western limb two weeks before, but often it takes a precise plot on both the first and second passages to pin it down for sure.

Another useful project is to plot the longitudes of the spots against time on a graph, as for the latitudes. You can do this for just one month or one solar rotation (27 days) or for several years. Using this technique, observers have found that some longitudes are sometimes more active than others.

Detailed Drawings

If you have time, it is sometimes worth making a more detailed drawing of an interesting spot or group at higher magnification. You can make such drawings using

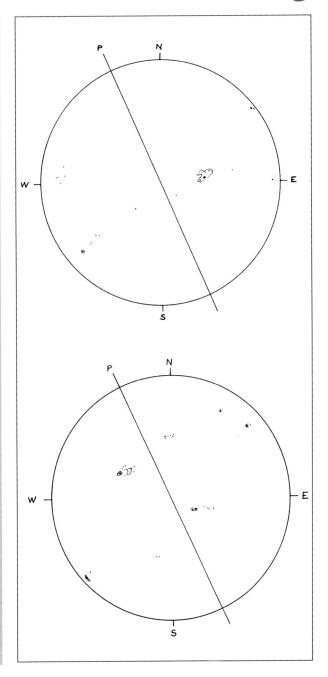

Figure 4.4. Two disc drawings made a week apart, showing the effect of the Sun's rotation on sunspot positions. The upper drawing was made on 2000 March 12th and shows a large D-class sunspot group in the southern hemisphere. A group in the northern hemisphere is appearing at the eastern limb. In the lower drawing, made on 2000 March 19th, the southern hemisphere group is now at the western limb, while the group seen in the east on the 12th is now just past the central meridian.

a grid as you would for a whole-disc plot, but with a blank sheet of paper placed over the grid on the clipboard. You could prepare a small grid specially designed for drawing individual sunspot groups, but I have always found it just as easy to project the high-power image onto the grid used for the whole-disc drawings and use that to draw the spots relative to each other. The essential thing to remember when making detailed drawings is to start by drawing the positions of the major sunspots on the grid squares first and add the finer details later. Do not try to draw in too much detail first, as the result will be less accurate. A well-aligned equatorial mount with an efficient motor drive is especially useful for making high-power drawings. Detailed drawings such as this have in some ways been superseded by photography, which can record the positions of sunspots relative to *each other* (as opposed to their positions relative to the points of the compass) more accurately than drawing. However, for many of us it offers a cheap and convenient way of recording sunspot detail. In any case, it is a good idea to know how to do it, as sometimes it is the *only* way in which to record an interesting feature before clouds roll in or a very rare phenomenon such as a white-light flare. White-light flares last for only a few minutes and could easily fade from view before you have set up your camera and filter. On the other hand, you can begin drawing such a feature pretty much as soon as you see it.

Co-operation with Other Observers

In an ideal world the amateur astronomer should observe the Sun every day, as only then can a complete picture of the Sun's behaviour be obtained. But in reality no one person can monitor the Sun every day. Depending on your local climate, some days will be cloudy, while the restrictions imposed by our daily lives can severely limit the time available for solar observation. In particular, if you live in a high temperate latitude such as the UK or parts of the US and Canada, you may well find that you cannot observe the Sun at all during weekdays in winter. For example, whereas I regularly manage to observe the Sun on 15–20 days during the summer months, I feel I have done well if I make observations on more than 5 days per month in winter. In any case, the Sun is by definition only visible during the daytime and so a constant watch on the Sun can only be maintained by observers scattered around the globe. However, you can help in the effort to monitor the Sun by sending in your observations to a central co-ordinating body, which pools together results obtained by a large number of observers and produces reports on the Sun's activity. Observing as part of a team applies not only to drawings and sunspot positions but also to sunspot counts and other observations described elsewhere in this book. Co-ordinating bodies for observations are often the solar sections of national amateur astronomy organisations and I have listed several such groups in Appendix B. I send observations to the solar sections of the Society for Popular Astronomy and the British Astronomical Association. The latter has over fifty regular observers operating from many countries, and for some years now the section has managed at least some coverage of the Sun on every single day of the calendar year.

Chapter 5

Measuring Solar Activity

The easiest and most useful type of solar observations an amateur astronomer can make is to measure the level of activity seen on the Sun. This is best done by counting the number of sunspots and/or sunspot groups visible on the Sun's disc each day. Other methods of measuring solar activity are used by professional observatories, but counting is the only one that is practical for the amateur. Sunspot counts are very useful in monitoring the Sun's behaviour and one type of sunspot count, known as the Relative Sunspot Number, is an internationally recognised measurement of sunspot activity.

Sunspot counting is quite easy and a simple daily sunspot count can take just a few minutes. This is good news if you are pressed for time, as while drawing the Sun is very satisfying to do, it can be quite time-consuming, especially if the Sun is active. For the same reason, sunspot counting is very suitable if you live in a cloudy climate, as in such conditions the Sun is often visible for only a short clear spell between clouds. Living in the UK, I know from personal experience how frustrating it is having to abort a drawing or photographic session half-way through because of clouds. For this reason, I always do my sunspot counting first during a solar observing session, because if clouds eventually roll in and curtail drawings, photographs and so on then I at least have something to show for my day's solar observing. Another major advantage of sunspot counting is that you can do it equally well whether you project the Sun's image or view it directly using a filter, since it does not require plotting or position measurements. It is therefore equally suitable for observers with Schmidt–Cassegrain or Maksutov telescopes as it is for those using traditional refractors and reflectors.

Try to make sunspot counts on as many clear days as you can. This is partly so that you can gain practice at the various methods of counting, but also because the more data you gather, the more accurate a picture of the ebb and flow of the sunspot numbers you will build up. As with drawings, assembling a complete picture of the Sun's behaviour is only possible if you make observations in collaboration with other amateurs, and I will describe methods of reporting results to national solar organisations below. Maintaining daily observations is especially important in periods of bad weather, as only a small number of observers may be active then and your results could be important. I have occasionally observed the Sun out of a window with light rain falling at the tail end of a shower, in order to get my daily sunspot number!

The Mean Daily Frequency

The simplest form of sunspot counting is to count the total number of sunspot groups – sometimes known as "active areas" – seen on the Sun's disc each day. At the end of the month, add up the numbers and then divide the result by the number of days on which you made sunspot counts. The resulting average is known as the *Mean Daily Frequency* of active areas.

To understand how the MDF works, let us use an example from my own observing notebook. Table 5.1 lists my counts of active areas made in May 2001. The left-hand column gives the day of the month on which an observation was made and on the right is the number of active areas seen on that day.

Adding up the figures on the right-hand column gives us 104. I made counts on 14 days in May 2001, so if we divide 104 by 14 we get an MDF of 7.4. This is quite a high MDF, indicating an active Sun – just what we would expect close to solar maximum.

Officially, any spot or group of spots which is at least ten degrees of heliographic latitude or longitude away from any other spot counts as one active area. The size

Table 5.1

Day	Active areas
4	8
5	8
6	6
8	5
12	6
17	6
21	6
23	10
24	7
25	9
26	10
28	9
29	7
31	7

or complexity of a group does not matter. Whether it is a single, small spot or a large, complex group, it counts as one active area. For a group to be counted, however, it *must* contain at least one sunspot with a dark umbra. An isolated area of faculae which does not contain any sunspots is not included in an active area count. Neither are pores (described in Chapter 3), isolated pieces of penumbra or the occasional greyish patches included.

Two exceptions to the above "ten-degree rule" occasionally occur. One is when a large group expands to the point where it is more than ten degrees long and contains two distinct centres of activity more than ten degrees apart. Such a group should then be counted as two active areas. The other is when two obviously different groups form less than ten degrees away from each other, and these should again be counted as two active areas. The ten-degree rule may make it seem difficult to determine what counts as an active area, but in practice the vast majority of groups are very distinct and separated by well over ten degrees. If a borderline case does occur, however, the best way of ascertaining whether it is one or two active areas is by making a disc drawing as described in the preceding chapter. Then, using the appropriate Stonyhurst Disc, it is easy to measure the separation between the groups and determine whether they are more than ten degrees apart.

When making an active area count, begin by counting all the groups you can see on the Sun's disc, using a low magnification. Then, switch to a high-power eyepiece and carefully scan the Sun in north–south strips, as you would for a drawing. Again, you will often be surprised at how many spots you pick up that were not visible at low power. Be especially vigilant when searching the east and west limbs of the Sun, and also when scanning the high-latitude zones, where spots are relatively rare but small ones occasionally appear. Because every isolated spot, however small, counts as an active area, a large number of small spots, undetected until you have scanned the Sun at high power, can make a big difference to your active area count.

If you see any exceptions to the ten-degree rule mentioned above, make a disc drawing. If you don't have time to make a drawing you must use your own judgement, but in any case you should always make a written note of your decision. You can, in fact, make a reliable active area count from a drawing, at your leisure after the observation, provided you have carefully examined the Sun at high power and recorded the positions of all the spots present. As I remarked above, however, it is often not safe to do it this way round in many climates, where clouds can roll in before you have completed your observation!

The Relative Sunspot Number

The standard method of counting sunspots used by astronomers throughout the world is somewhat more sophisticated. Known as the *Relative Sunspot Number*, it takes into account both the number of sunspot groups (active areas) present and the number of individual sunspots within these groups. The Relative Sunspot Number was started as long ago as 1848 by Rudolf Wolf, director of the Swiss Federal Observatory in Zurich, and has been in use ever since. Nowadays the central co-ordinating body for the Relative Sunspot Number is the Sunspot Index Data Centre in Uccle, Belgium, where an official sunspot number for each day is worked out using results obtained by an international network of observers. A similar system operates in the United States to produce an American Relative Sunspot Number. Amateur solar observing groups worldwide send their results to both organisations. It gives the aspiring solar observer some encouragement to know that Wolf made his original observations with an

80 mm f/14 refractor – exactly the sort of instrument employed by modern amateur observers – and, even more remarkably, that the same telescope is still used daily to count the sunspot number! One of the great strengths of the Relative Sunspot Number is that it continues an unbroken record of solar observations, made with the same type of instrument, going back over 150 years.

The daily Relative Sunspot Number, usually abbreviated to R, is worked out using the following formula:

$$R = k(10g + f)$$

where g is the total number of sunspot groups (active areas), counted in exactly the same way as for the MDF and f is the total number of individual sunspots within the groups. k is a constant included to standardise results from a wide variety of observations. Sunspot counts vary widely between observers for many reasons, including the size, type and quality of the telescope, local seeing conditions and the ability and experience of the observer. Determining your own constant requires making sunspot counts over a long period and then comparing your results to the official, published values for the same period. However, solar observing organisations often do this for you and so you can assume, to begin with, that R is simply $10g + f$.

Let us use a few examples to show how R is determined. Suppose we see three active areas on the Sun. One large group contains 20 individual spots, the second contains 8 and the third is just a single, isolated spot. The total number of individual spots f is therefore $20 + 8 + 1 = 29$. We need to add this figure to the total number of groups g multiplied by 10, i.e. $3 \times 10 = 30$. Therefore our final R is $30 + 29 = 59$. The smallest possible value for R that is not zero is when there is just one single spot on the entire solar disc. R is then 11 – i.e. 1 group + 1 spot. To choose a slightly more active example, Table 5.1 shows that on 2001 May 4th I saw 8 active areas. Some of these contained a large number of spots and I counted a total of 70 individual spots. The resulting R was therefore $80 + 70 = 150$. One of the highest levels of activity I have ever seen was on 2000 July 19th, when there were no less than 16 groups and 193 individual spots, giving an R of 353.

To do the sunspot counts to enable you to calculate R, begin by counting the number of active areas exactly as you would for an MDF count, remembering to scan the Sun at high power for small spots. Then, again using the high-power eyepiece, count all the spots that you can see within the groups. Always count the groups first and then the spots, for the same reason as it is advisable to do sunspot counts before drawings: if during the spot count it clouds over for the rest of the day, you have at least got a basic MDF count.

To count the number of spots correctly it is important to know what is officially a spot and what is not. To be included in a count a spot must contain an umbra and must be entirely separate from any other spots in a group. If you see what looks like two spots joined together, always count them as just one spot unless they are entirely separated by a bright area between them. If a light bridge appears in a spot and divides it into two, then you should count it as two spots – but only when it is completely divided. A spot with a light bridge running only part of the way across it is still one spot – although you should make a note that a light bridge is developing. As with MDF observations, you should not include pores and penumbrae in spot counts. When a group is near the east or west limb, or when seeing conditions are less than good, it is often difficult to distinguish pores and small sections of penumbrae from genuine umbrae. Small pieces of penumbra are often found around the edges of large, complex groups, particularly when they are starting to decay. When deciding whether a feature is a genuine spot, compare how dark it appears with nearby "real" sunspots. A genuine umbra appears very dark, while penumbrae are a lighter grey when examined closely.

At times of high activity, when there are groups containing large numbers of individual spots f – the total number of spots on the disc – can sometimes be 100 or more, and it is easy to lose count when tallying them up. One way to get round this is to note down the number of spots in each group and then add up the total at the end of the observation. In any case, it is often a good idea to count f twice, if possible, in order to check that you have not made any major errors in counting.

When you have completed the sunspot count, do not forget to record the time at which you finished the observation (in Universal Time, to the nearest minute) and the seeing conditions, as you would for a drawing. It is also important to note down full details of the telescope used, including whether you used projection or observed the Sun directly with a filter. If you observe the Sun directly, you should record the magnifications of the eyepieces used, while if you use the projection method you should note down the approximate diameter of the Sun's disc as seen on the projected image. You can get the disc diameter for a high-magnification projected image by dividing the magnification of the high-power eyepiece by that of the lower one, and then multiplying the result by the diameter of the whole-disc image used for low-magnification work. For example, if you use an eyepiece giving a magnification of 56× to produce a 152 mm (6 inch) solar image, then the disc diameter produced by a 7 mm (100×) eyepiece used on the same telescope is $(100 \div 56) \times 152 = 271.4$ mm (approximately 10.7 inches). If you stick with the same instrument and projection setup, then you need only record your instrument details once, but always record any changes you make to your instrument.

For a slightly more detailed report, you can divide your counts of both active areas and R into hemispheres. The northern and southern solar hemispheres often differ widely in the level of activity over the course of a month and sometimes over longer spells, and amateur solar observing groups publish MDF and R figures for each hemisphere as well as general data for the whole Sun.

Knowing what is the true northern and southern hemisphere on the Sun requires a knowledge of P (the angle of the Sun's axis to true north) and B_0, as you would need when making position measurements from drawings. While it is always best to make a drawing if you have the time, you do not always need to draw the disc to establish the positions of the hemispheres for sunspot counting purposes. Because sunspots tend to occur in distinct bands of latitude in each hemisphere, it is often obvious which spots are in the northern hemisphere and which are southern, provided the image is correctly orientated as described in Chapter 4 and you know the value of P. Occasionally, however, spots do form close to the equator, at latitudes of 5° or even less, and you will then need to make a drawing to be certain which hemisphere they are in. This often occurs around solar minimum and the years immediately preceding it when, in accordance with Spörer's Law, sunspots of the dying solar cycle form at lower latitudes.

To establish the orientation of the Sun's axis, look up the value of P in the *BAA Handbook* or the *Astronomical Almanac*. Once you have established the position of apparent north by letting a sunspot drift along the east–west line, imagine that the north–south line is tilted by the appropriate value of P. The east–west line is tilted by the same amount, and this roughly marks the division between the hemispheres. To establish the angle with any degree of precision, you need to mark the circumference of the disc on your projection grid with angles at $5°$ intervals. Unless you are observing during the late stages of a solar cycle, it should now be clear which hemisphere most of the spots are in. You may notice, however, that a few spots seem to pass through the exact centre of the disc – an unlikely occurrence, given that spots exactly on the equator are rare. This is due to the Sun's tilt towards us, as determined by the value of B_0. If B_0 is positive, the Sun's north pole is tilted towards us, and so the equator curves below the centre of the disc, while if it is negative the equator curves above. You need to take this into account when assessing which hemisphere a spot is located in. The business of assessing the Sun's true orientation by eye may sound rather obscure, but I have found it to be quite easy after a little practice, using the projection method (Figure 5.1). However, if you are still not confident about establishing north and south without making a drawing, you can leave this out of your sunspot counts, as the essential statistics are the total counts of active areas and spots on the whole solar disc.

Recording and Reporting of Results

It is important that you record your sunspot counts in a standard format, as this ensures that you record all the correct details. It also allows you to easily transfer your results onto the report forms provided by solar observing groups and enables you to compare your own observations directly with the results published by these groups. The best way of recording sunspot counts is in a table for each month's observations, with a row for each day of the month, and Figure 5.2 shows a sample page from my own solar notebook. I prepared my standard table on a computer using Microsoft Word software and I paste a fresh pre-printed table in my notebook for each month, but you can draw one out by

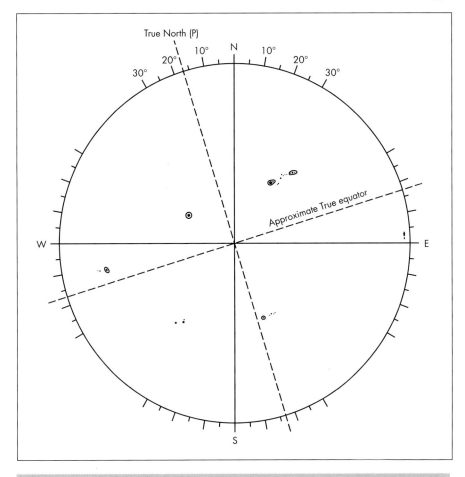

Figure 5.1. Diagram showing how to determine the Sun's true orientation by eye. Notice that the spot group in the west is in the apparent southern half of the disc, but when the Sun's true orientation is taken into account it is really in the northern hemisphere. Note also that the Sun's true equator is usually curved northwards or southwards and this needs to be taken into account when determining which hemisphere a spot is in.

hand just as easily. The headings of the columns are as follows:

D   Day of the month
UT Universal Time at which the sunspot count was completed
S   Seeing (on the 1–5 scale described in Chapter 4)
g_n Number of active areas seen in northern solar hemisphere
f_n Number of individual spots seen in northern solar hemisphere

SUNSPOT COUNT (White Light)　　June　　　　2001

Day	UT	S	g_n	f_n	g_s	f_s	g	f	R
1									
2									
3	17:45	3	3	26	3	29	6	55	115
4	18:59	3	4	39	4	31	8	70	150
5									
6									
7	19:49	5	6	27	3	25	9	52	142
8	17:12	4	5		3		8		
9	19:19	3	6		3		9		
10	18:46	3	7	59	4	33	11	92	202
11									
12									
13	18:58	3	7	62	4	12	11	74	184
14									
15	19:02	4	7	65	4	22	11	87	197
16									
17									
18									
19	18:55	3	4	65	3	13	7	78	148
20	19:22	3	3	85	4	28	7	113	183
21	19:27	5	3		4		7		
22	18:45	2	3	74	4	35	7	109	179
23	19:36	3	3	55	4	25	7	80	150
24	18:56	2	3	56	4	32	7	88	158
25	18:54	2	3	48	4	45	7	93	163
26									
27	19:09	4	2	26	5	26	7	52	122
28	19:28	4	2	10	4	11	6	21	81
29									
30	17:02	4	5	10	4	14	9	24	114
31									
TOTALS			76	707	68	381	144	1088	2288
AVERAGES			4·22	■	3·78	■	8·0	■	152·5

MDF = ...8·0........　　R = ...152·5........　　Obs. Days = ...18........

g_s Number of active areas seen in southern solar hemisphere

f_s Number of individual spots seen in southern solar hemisphere

g Total number of active areas seen (found by adding g_n to g_s)

f Total number of individual spots seen (found by adding f_n to f_s)

R Relative Sunspot Number, i.e. $10g + f$

There are two additional rows at the bottom of the table. The first is the total for each column, added up at the end of the month. This facilitates finding the average number of active areas seen (i.e. the MDF) and Relative Sunspot Number, which I write in the final row. For quick reference I also find it helpful to make a note at the bottom of the page of the month's MDF, R and the number of days on which I made observations. A sunspot count table does not have to be as elaborate as this. I divide my count up into hemispheres, which increases the number of columns, but if you wish to do sunspot counts for just the whole disc then you only need five columns: two for g and R, in addition to those for the day of the month, the time and the seeing conditions.

In addition to recording the figures, I like to make a few brief notes on what I have seen each day, especially if I have seen something interesting or unusual – a new large group at the eastern limb, a spot split by a light bridge or an especially complex or active sunspot group, for example. Such notes are especially useful as an aid to my memory if I did not make a drawing or was unable to observe the Sun for several days afterwards. They can also be important in their own right. For instance, at 12:45 UT on 1992 February 15th, I noticed a fairly small spot near the centre of the Sun's disc whose umbra appeared to be broken into several fragments. I did not make a drawing that day, but made a note of the spot's unusual appearance and included it with my monthly report to the BAA Solar Section. Some weeks later, I read in the Solar Section's monthly newsletter that several other observers had noticed unusual activity in the spot and that the following day the same spot had produced a white-light flare – an extremely rare event, as we have already seen. If only I had observed the Sun the next day as well! Nevertheless, this event shows how important it is to make notes, as you never know how useful they might be.

Figure 5.2 (*opposite*). Sample page from the author's solar notebook, showing sunspot counts recorded in tabular form.

It is important to send your sunspot data to your solar observing organisation promptly after the end of the month, so that they can collate and publish the results obtained by many observers as quickly as possible. The quickest way of reporting is, of course, by e-mail, although you must ensure that the file attachment containing your report is in a format that the group co-ordinator's computer can read. If it is not in the correct format the co-ordinator may not be able to read it or – worse – it may contain "rogue" characters that could confuse the report. Personally, I still prefer to send in my observations by post. If you submit observations in this manner you should use one of your group's standard report forms, as it is easier for the co-ordinator to collate and analyse observations if they are in the same format. The table in my solar notebook is deliberately drawn up to closely resemble the report forms issued by the Solar Sections of the British Astronomical Association and Society for Popular Astronomy, to both of which I send monthly sunspot counts, brief notes on my observations and, if possible, drawings.

Whether you just count active areas or measure R as well, it is a fascinating exercise to plot a graph of solar activity against time. To begin with, you can do this for a single year, plotting the number of active areas or R against the months on a simple line graph. You can plot graphs using graph paper, if you wish, but nowadays it is also possible to draw them on a computer, using software like Microsoft Graph, which comes with modern versions of Microsoft Word. Unless the solar cycle is close to maximum or minimum, the resulting graph will almost certainly show an upward or downward trend, depending on which part of the cycle the Sun is in. Monitoring sunspot activity levels over several years produces an even more interesting graph. Figs. 5.3 and 5.4 show graphs of the MDF and R respectively, plotted from my own sunspot counts. Both graphs begin in 1996, shortly before the minimum of solar cycle 22, and end five years later. Graphs plotted over several years have a jagged appearance, as activity can vary greatly from month to month, but they clearly show activity bottoming out around 1996 and rising towards the next peak at the end of the decade. The shape of both graphs agrees quite well with the graphs produced by amateur solar groups from observations by many workers. Many peaks and troughs in the graph agree as well – for example, in the MDF graph the great peak of July 2000 and the lesser peak of June/July 1999 both show up in the collective graph, as

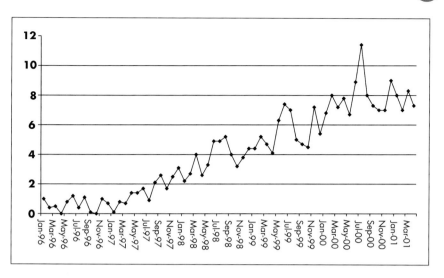

Figure 5.3. Graph plotted from the author's sunspot counts, showing the variation in the MDF between 1996 and 2001.

does the trough of August and September 1999. But results can easily be distorted if you do not make many observations in a month. This can often happen in winter, when the few days on which you manage to make observations coincide with a period of unusually high or low activity. This gives an erroneous MDF or R figure for the month and causes a peak or trough on your personal graph which is absent on the collective graph. An example of this happened to me in March 2001. This was a very cloudy month and there were only three days in which I was able to count the Relative Sunspot Number. Two of these clear days were at the end of the month, the time at which the very large sunspot group pictured in Figure 3.4 (Chapter 3) was on the disc. This group contained many spots, giving very high R counts on both days. Activity earlier in the month, however, was much lower, and so my average R for the month was higher than it should have been. The peak visible in Figure 5.4 did not show up in the collective graphs plotted by solar observing organisations.

Graphs can be used to tell at a glance when the maximum and minimum of the solar cycle passed – but only after the event. It is obvious from both graphs that the Sun reached minimum activity in late 1996. We might also conclude that sunspot maximum occurred with the great peak in activity of mid-2000, because as of the time of writing (late 2001), sunspot activity has not reached this level again. But we must be cautious when making assessments so close to maximum or minimum, because the Sun can, and does, frequently surprise us. At

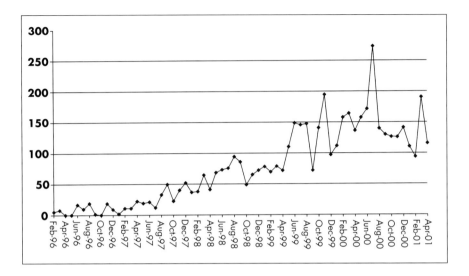

the maximum of cycle 22, for example, sunspot activity reached a high peak in mid-1989 and then subsided, leading some astronomers to speculate that 1989 was the date of maximum. But activity flared up again in 1990 and once more in late 1991. Only by plotting activity over several years around the time of maximum can we be sure that maximum has really passed.

Figure 5.4. Graph plotted from the author's sunspot counts, showing the variation in the Relative Sunspot Number *R* between 1996 and 2001.

Observing Faculae and White-Light Flares

As well as sunspots, you can also keep records of faculae. However, as explained previously, faculae are only easy to see when they are close to the limb. The best way of recording them is by drawing their positions and outlines on whole-disc drawings, in the same way as for sunspots, or by photographing them on high-contrast film or digital media.

The most interesting type of faculae are the polar faculae, partly because of the challenge of seeing them and also because relatively little is known about them. To see these faculae you need a high-contrast solar image viewed at high magnification in good seeing. Contrast is especially important, as polar faculae are low-contrast features, and if any daylight is allowed to fall onto the image your chances of seeing them are much reduced. I use my 130× eyepiece to project a 260 mm (10½ inch) solar disc in my projection box, which gives good contrast by shielding the solar image from most of the daylight. In fact, I have usually had my best views of polar faculae when observing the Sun's image in the projection box inside a curtained room, the curtains providing additional blackout. If you observe the Sun directly with a filter, use a rubber eye guard on the eyepiece to prevent stray light from reaching your eye. Some observers even shroud their entire face with a black cloth, as old-time photographers used to do when changing the plates.

To find polar faculae you need to know the values of P and B_0 for the day you are observing. Faculae are always easiest to see at the pole that is tipped towards the Earth. Therefore the best time to see faculae at the north pole is in autumn, as this pole reaches its maximum tilt towards us ($B_0 = +7.2°$) in September. Similarly, southern polar faculae are best seen in spring, B_0 reaching $-7.2°$ in March. In midsummer and midwinter, when B_0 passes through zero, it is sometimes possible to see faculae at both poles. How many polar faculae you will see at either pole depends on the current stage of the solar cycle, as their numbers are highest at solar minimum and the declining phase of the cycle just before minimum.

To look for polar faculae, start at low power and ensure that the whole-disc image is correctly oriented using the spot-drift method. Then read off the value of P for the pole currently tipped towards Earth on the angular scale at the edge of the projection grid. This is the part of the limb you should next examine with the high-power eyepiece. Remember that polar faculae are not large patches or streaks like spot-related faculae, but rather are very small, bright spots, sometimes little more than points of light. In addition to the contrast-enhancing measures mentioned above, you may find the faculae easier to see if you move the telescope very slightly back and forth using the slow motion controls. This is because the human eye sometimes sees low-contrast objects better when they are moving (indeed, the same trick is used by deep-sky observers when searching for faint objects). Also, if you use the projection method, moving the image is a good way of assessing whether polar faculae are real and not just irregularities on the surface of the paper. Regardless of the value of B_0, carefully scrutinise both poles for faculae, as while faculae are harder to see and much less frequent at the pole tipped away from Earth, one or two are still sometimes visible. If you see any very bright or unusually shaped faculae at the poles, make a note of them and do a basic sketch.

The easiest way of making useful records of polar faculae is to simply note whether any are present each day you observe the Sun. Record at which pole or poles they are visible and also make a note of the seeing conditions at the time you are observing. At the end of the month you can calculate the percentage of observing days on which you saw polar faculae and plot this figure on a graph against time. In fact, you could plot this on the same graph as, say, R, with a percentage scale on the opposite side to the Y-axis, to show how the frequency of polar faculae varies over the years. To make the statistics more accurate, you could include in your calculations only the days in which the seeing was at least reasonably good. Counting the numbers of individual polar faculae is very difficult, because the faculae are so small and elusive that they tend to drift in and out of visibility due to atmospheric turbulence and their positions are difficult to memorise.

If you are ever fortunate enough to see a white-light flare, you should report it immediately to your observing group co-ordinator by the quickest method possible, preferably by telephone. However, if you are relatively new to solar observing, contact an experienced observer at your local astronomical society and, if possible, get him or her to verify your observation. As I said in Chapter 3, it is easy to mistake a light bridge or other types of bright area within a sunspot for a flare, and so there is the danger of causing a false alarm. If you see what you believe to be a white-light flare, record the time to the nearest minute at which you first noticed it, the time at which it reached maximum brightness and the time at which it faded from view. You should also, of course, record the details that you would for any solar observation, such as seeing conditions, instrument, method of observation and so on. Make a basic sketch of the flare in relation to its parent sunspot group and, if possible, photograph it using the methods presented in Chapters 7 and 8.

Observing Naked-Eye Sunspots

A large sunspot group is sometimes visible to the naked eye, provided your eye is protected by a safe solar filter. Filters used for telescopic solar observing, such as Mylar or glass aperture filters, are just as effective for observing the Sun with the naked eye and, as mentioned in Chapter 2, a welder's glass is also suitable, provided it is of shade number 14. Whichever type of filter you choose, try to obtain one that is large enough for the Sun to be seen with both eyes at the same time. This is more comfortable than squinting with just one eye and you will be able to see any naked-eye spots more clearly if your eyes are relaxed.

Observers have sometimes seen naked-eye sunspots without a filter, when the Sun is setting or is dimmed by mist or thin cloud. Indeed, in Chapter 1 I mentioned that the earliest recorded observations of sunspots were made in this manner by the ancient Chinese. However, while it is worth noting if you see a spot in this way by accident, I do *not* recommend it as a method of deliberately looking for naked-eye spots. You can never be sure that the amount of solar radiation getting through to your eyes is small enough not to be dangerous.

A sunspot – or, more usually, a sunspot group – has to be at least fairly large to be visible to the naked eye. A rough rule of thumb used by astronomers for many years is that a group has to be at least three times the size of the Earth in its longest dimension for it to be visible. But we have to take several important factors into consideration. For example, a spot must be well away from the limb for it to be seen with the naked eye, as when close to the limb its apparent size is greatly reduced by the effect of perspective. Also, the compactness of the group needs to be taken into consideration. A group containing large areas of umbra in a compact arrangement is much easier to see than a group of the same size whose spots are sparsely scattered. Finally, there is the observer's eyesight. An observer with perfect or near-perfect long-range vision can see smaller sunspots than someone with mediocre eyesight.

It is interesting to check each day whether any naked-eye sunspots are visible and some solar observing organisations keep systematic records of naked eye sunspots as a way of monitoring solar activity. To do this, examine the Sun with your filter once each day and note down the number of sunspots you can see. Always look for naked-eye sunspots *before* observing the Sun with a telescope. If you see a large sunspot with your telescope first, your chances of seeing it unaided will be much higher, because you will know it is there and will know its position on the Sun's disc as well. However, this observer bias is sometimes unavoidable, as you may well know that a large spot is present because you saw it using the telescope the previous day. Usually the number of naked-eye sunspots seen each day will be 0 or 1. At solar minimum, looking for naked-eye sunspots can be patience-testing, as there are periods of weeks or months when no naked-eye groups are visible. Near maximum, however, naked-eye spots are quite common and there are sometimes days when two or more spots are visible on the disc at the same time. At the end of each month, add up the number of spots and average them to get an MDF, as for sunspots seen using the telescope. You can then plot the MDFs on a graph to show how naked-eye sunspot numbers vary over time. Past observations indicate that naked-eye sunspot numbers correspond quite closely with the MDF and R as seen through telescopes, so that the maximum number of naked-eye sunspots occurs roughly at the time of telescopic sunspot maximum.

Chapter 6

Observing the Sun in Hydrogen-Alpha

In Chapter 1 we learned that the Sun is surrounded by an extensive, invisible atmosphere. This is in two parts: the chromosphere, a thin layer a few thousand kilometres thick, and the corona, which consists of rarefied but extremely hot gas extending millions of kilometres out into space. Both layers are invisible to the Earth-bound observer in ordinary circumstances, because the brilliant light of the photosphere is scattered by the Earth's atmosphere and so washes out the relatively feeble light of the solar atmosphere. The only time an Earth-based observer can see the Sun's atmosphere without special equipment is during the fleeting few minutes of a total solar eclipse. To see the corona from Earth without waiting for an eclipse is very difficult, due to its extreme faintness relative to the Sun's surface. Professional astronomers can photograph the corona using an instrument called a *coronagraph*, a telescope containing an occulting disc to precisely block out the Sun's surface, and optics specially made to keep scattered light from the photosphere to an absolute minimum. For a coronagraph to work, it must be installed at a very high mountaintop site, where scattering of light by our own atmosphere is minimised. Even then, a coronagraph can only image the inner parts of the corona. The best images of the corona are obtained from space-based solar observatories.

The chromosphere, however, is a different matter. Anyone familiar with basic astronomy will know that light from the Sun and other stars is made up of a spectrum of colours, running from red to violet. Unlike the photosphere or the corona, whose spectra are continuous streams of colours, the chromosphere emits most of its light in a number of fine lines, each of them covering a very narrow band of the visible spectrum. If we can isolate one of these lines and observe the Sun in just this wavelength, most of the brilliant light of the Sun – and consequently the scattering by the Earth's atmosphere – would be eliminated, and we would see just the light of the chromosphere. The brightest line emitted by the chromosphere is known as hydrogen-alpha (abbreviated H-alpha or Hα) and is located in the red part of the spectrum. The chromosphere also emits in hydrogen-beta, in the blue region of the spectrum, as well as helium (yellow) and calcium (blue). All these wavelengths – and many more – are studied by professional astronomers, but H-alpha is much the easiest to observe and most instruments made for the amateur are designed to show the Sun in this wavelength.

Astronomers use two basic types of instrument to isolate the H-alpha spectral line. The oldest method involves splitting up the solar spectrum to a sufficient extent that an individual emission line is resolved and isolated. A more recent development is a special type of filter known as an interference filter, which cancels out all wavelengths of sunlight reaching the observer except H-alpha. These filters are nowadays available commercially, at prices amateurs can afford. It is wonderful that the amateur can see the chromosphere, for in this layer we can see many features that are not observable in white light and we can begin to see some of the activity revealed so dramatically in images taken by professional and space-based observatories.

When we observe in H-alpha (Figure 6.1), the Sun's disc has a brilliant red colour and the chromosphere appears as a thin, faint band around the limb. The "surface" – i.e. the chromosphere – has a granular structure, but this is much larger and easier to see than the photosphere granulation seen in white light. Sunspots are still visible, but they are not as prominent as they are in white light, because we are looking at them through a welter of activity in the chromosphere. Four major solar activity features are visible, all of them associated with magnetic fields and broadly following the solar cycle in their numbers and intensity. The most dramatic H-alpha features are the *prominences* – large, flame-like structures of hydrogen at the limb of the Sun

(Figure 6.2). Many of them are tens of thousands of kilometres high and project out into what looks like the blackness of space but which is, in fact, the invisible corona. Prominences vary hugely in size, shape and structure. Some appear as small spikes at the limb of the Sun, others are intricate loops appearing exactly like the arrangement of iron filings on paper above a bar magnet. Another type extends to only a low height in the solar atmosphere but covers a very large area of the limb. Some types appear to erupt out of the Sun like lava from a volcano and change structure very rapidly. Changes in prominences do not normally occur fast enough for them to take place before your very eye, because at the Sun's distance material being ejected at very high speeds appears to be moving quite slowly. Prominences are often associated with sunspots, but not always, and sometimes when sunspot activity is quite low it is surprising how many prominences are visible around the limb in H-alpha.

If you observe the Sun with a suitable H-alpha instrument on almost any day you will probably notice at least one or two sinuous, dark lines on the disc which are not visible in white light. These are called *filaments* and are simply prominences seen silhouetted against the disc. It is sometimes possible to see their true nature yourself when a filament close to the limb extends into the darkness above the limb and shows up as a prominence. Like prominences, filaments are often associated with sunspots and sometimes appear to be entangled with sunspot groups, but on many occasions they occur well away from any spots and at high latitudes, where spots seldom or never form.

If there is any activity at all on the Sun, you will quite likely notice some large, bright patches, many of them surrounding sunspots. These are known as *plages* or *flocculi* and resemble the faculae seen around sunspots in white light, except that unlike the latter they are visible at the centre of the disc and not just when near the limb. Like the faculae, they often form several days before the sunspots themselves and linger for some time after the spots have disappeared. Many plages cover a considerable area and are a graphic illustration that a sunspot group is often just the "tip of the iceberg" in a large active region.

Finally, with an H-alpha filter solar flares become quite common features, several a day sometimes being visible at solar maximum. A flare shows up as a brilliant white patch which suddenly appears in the space of a few minutes. Although flares reach their peak brightness very quickly, they can take an hour or more to fade from view and can sometimes resurge again, extending the spectacle for much longer.

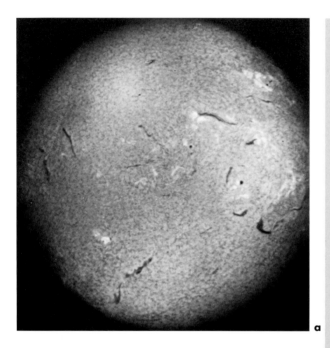

a

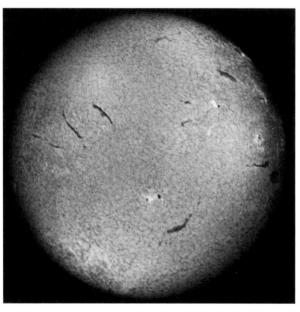

b

Figure 6.1. a The Sun's disc in hydrogen–alpha. Both photographs by Eric Strach, using a 200 mm (8 inch) SCT stopped down to 63 mm (2.5 inch), a DayStar filter of 0.6 angstrom passband and 1/90 second exposures on Kodak TP 2415 film. The dark linear features are filaments (i.e. prominences seen silhouetted against the disc), and the bright regions are plages. **a**: 2000 August 20th. **b**: 2000 August 22nd. Note the change in the position of the filaments due to the Sun's rotation.

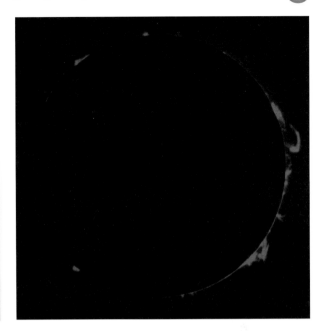

Figure 6.2.
Prominences around the whole Sun, photographed by Derek Hatch using a Baader coronagraph attached to an Astro Physics 150 mm (6 inch) f/9 refractor and a 2× teleconverter. Exposure 1/60 second on Fuji Sensia 400 film in a Mamiya 645 (medium format) camera.

Equipment for H-Alpha Observing

Unfortunately, the equipment required to isolate the H-alpha line and reveal the chromosphere is expensive. A typical H-alpha filter costs as much as a very good telescope, and this is the reason why only a minority of amateur astronomers observe in H-alpha. However, in recent years technological advances have brought the price of these filters down considerably. While a suitable filter remains a substantial investment, it is by no means outside the budget of many amateurs, especially when we consider the large sums many amateurs spend on CCD cameras and computer-controlled telescopes. Moreover, just as in white light, you can do H-alpha work with a small telescope.

So that we can understand how H-alpha filters work and assess what is the most suitable equipment to buy, I feel it is helpful to briefly look at the history of H-alpha solar observing. The first observation of prominences outside a total solar eclipse was made independently in 1868 by the British astronomer Sir Norman Lockyer and the Frenchman Jules Janssen, by aligning the slit of the recently invented spectroscope at a tangent to the limb of the Sun and thus isolating the

H-alpha spectral line. By opening up the spectroscope slit a little it was possible to see a prominence, if one was present at that part of the limb. Observing prominences in this way was rather cumbersome, because it meant viewing prominences one at a time by patiently scanning around the limb. But until the advent of affordable H-alpha filters almost a century later, the prominence spectroscope remained the standard method of viewing the prominences for amateur astronomers – and even those that had spectroscopes were a privileged few.

In the 1890s an improvement on the spectroscope, known as the *spectroheliograph*, was invented by the American George Ellery Hale and another Frenchman, Henri Deslandres. This allowed astronomers to photo-graph larger areas of the Sun, including disc features in the chromosphere such as filaments and flares. In the 1920s Hale made a modified version of this instrument, known as the *spectrohelioscope*, which allowed the disc features to be viewed visually. The spectroheliograph remains a standard instrument in professional observatories to this day and a small number of amateur astronomers observe the Sun using home-made spectrohelioscopes. The spectrohelioscope has a major advantage over H-alpha filters in that it can be "tuned" to any wavelength in the spectrum and thus show the chromosphere in any of the emission lines, including calcium and hydrogen-beta as well as H-alpha. An observer with a spectrohelioscope can thus compare the view of chromospheric phenomena in different wavelengths. However, spectrohelioscopes are not available commercially and require a great deal of optical and mechanical skill to build, putting them outside the reach of most amateurs. In addition, unlike filters they are very bulky and are not portable.

The development of filters for H-alpha observing began in 1930, when yet another Frenchman, Bernard Lyot, invented the coronagraph and took the first non-eclipse pictures of the corona from the high-altitude site of the Pic du Midi Observatory in the French Pyrenees. Later in the 1930s Lyot improved the coronagraph by incorporating an interference filter – a complex filter composed of many layers of quartz and polarising materials which, using the principle of interference, cancelled out all wavelengths of solar radiation except the H-alpha line. This enabled prominences to be imaged through the coronagraph, and interference filters were later improved to the point where they could show disc features as well.

To show H-alpha features at all, a filter (or spectro-helioscope) must isolate a tiny slice of the Sun's spectrum. Light at the red end of the spectrum has a wavelength of around 7,000 angstroms (700 nano-metres), while at the violet end the wavelength is about 4,000 angstroms. The visible spectrum, therefore, covers about 3,000 angstroms, but the H-alpha line is only about 1 angstrom wide. The extent of the spectrum allowed through by a filter is known as the filter's *passband*. The narrower the passband, the more the H-alpha line is isolated from the main part of the Sun's light and so the greater the contrast of the resulting images. Filters with narrower passbands are more complex, and therefore more costly, to make. A filter with a passband of several angstroms will reveal prominences, provided it is centred on the H-alpha line, the Sun is hidden by an occulting disc and the sky is free from contrast-reducing haze. But to reveal features on the disc it is necessary to use a "sub-angstrom" filter – i.e. one with a passband of less than 1 angstrom, narrower than the H-alpha line itself. Filters of the type employed by Lyot, using quartz and polaroids to isolate the H-alpha line, are still used by professional astronomers today as a standard means of imaging the chromosphere, but they are too bulky to fit on amateur telescopes and their five- or six-figure price tags put them well beyond the amateur's budget.

In the 1960s a new type of interference filter became available. Instead of quartz and polaroids it used a large number of layers of dielectric materials such as magnesium fluoride deposited on glass, known as an "etalon", to isolate the H-alpha line. These filters were much more compact and, most importantly of all, were available "off the shelf" at prices amateurs could afford. Initially they were only capable of showing prominences, because their passbands were wider than 1 angstrom, but a number of amateurs successfully developed prominence viewing devices using the new filters mounted in telescopes built on the coronagraph principle, with a disc to occult the Sun. The following decade the US-based DayStar Filter Corporation devel-oped sub-angstrom filters of the same type, enabling amateurs for the first time to observe disc features as well as prominences with a filter attached to an ordinary telescope.

Because the market for such specialised filters is very small, only a handful of companies around the world currently produce them. Nevertheless, the amateur astronomer today can choose from a number of commercially available H-alpha filters, both sub-angstrom filters and those capable of showing just prominences. All are expensive by ordinary standards, but they are a good investment, considering the wealth of otherwise invisible solar detail they reveal.

For an interference filter to work, the light rays striking it must be close to parallel. In most amateur telescopes, whose f/ratios are rarely longer than f/15 and often much shorter, the light rays are strongly convergent, which has the effect of widening the passband of the filter and so reducing the amount of detail visible through it. H-alpha filters work best if the effective f/ratio of the telescope is increased to about f/30, so that the light rays are closer to parallel. With many H-alpha filters, you need to use a *pre-filter* or *energy rejection filter* (ERF) mounted in front of the telescope's aperture to achieve this. The ERF increases the f/ratio by stopping down the aperture. It also contains a filter made of heat-resistant red glass to protect the main H-alpha filter (and your eyes) from the solar heat. Never use an H-alpha filter without an ERF: you could risk damaging your eyesight, as well as the filter. Because the size of the ERF varies with the aperture and f/ratio of your telescope, you need to specify these details when ordering an H-alpha filter.

The US firms Lumicon and Thousand Oaks Optical both sell H-alpha filters with passbands of 1.5 angstroms. As they are not sub-angstrom filters they are generally for prominences only. However, they have no occulting disc to blot out the bright Sun, as the filter dims the image enough for the disc to be viewed safely. When skies are transparent, it is possible to see some of the larger and darker filaments and even the occasional flare through these filters. Both the Lumicon and Thousand Oaks models use ERFs and both are priced at around $800, including the ERF. I do not recommend them for use with very small telescopes, as the aperture stop required to reach f/30 would give a very low resolution. For example, to achieve f/30 on a 60 mm f/15 refractor would require an ERF with an aperture of only 30 mm! However, when used with medium-sized or larger telescopes (including SCT and Maksutov instruments) these filters are easy to use and their affordable prices make them very suitable for the newcomer to H-alpha observing.

If you are content to view just prominences, and want the best possible views of them, the H-alpha

Coronagraph manufactured by Baader Planetarium (who also make the AstroSolar Safety Film for white-light observing) is an excellent choice (Figure 6.3). This device screws onto an ordinary refractor and works on the principle of the coronagraph, using a specially made metal cone to produce an artificial eclipse of the Sun's disc. But this device shows only prominences, not the corona. At its heart is an H-alpha filter with a 1.5 angstrom passband – again, too wide for most disc features, but more than narrow enough to reveal prominences. Unlike other H-alpha filters, when used with telescopes of 80 mm aperture or smaller the Baader coronagraph does not require an ERF to be fitted at the front of the telescope. Instead, the system's effective f/ratio is increased by special lenses inside the device. The beauty of this is that the telescope can be used at *full aperture*, allowing prominences to be seen in exquisite detail, the resolution limited only by the aperture of the telescope and the seeing conditions. Because the Sun's surface is masked by the occulting disc, the prominences appear extremely bright. The effect really is like a total solar eclipse – although without the corona, of course. The coronagraph, which costs around £1,100 ($1,500), can be purchased with a device allowing individual parts of the limb to be viewed at high magnification. It can be used with an ordinary $1\frac{1}{4}$ inch star diagonal and eyepiece and it also contains a standard thread for a camera, enabling any camera with a suitable adapter to be attached for photographing the prominences.

The system does, however, have some disadvantages. First, it is less compact than other filter types, the

Figure 6.3. Baader coronagraph on the author's 80 mm (3.1 inch) refractor.

version for an 80 mm refractor adding some 200 mm (8 inches) to the length of your tube. It can also be inconvenient in that it consists of three parts, which need to be threaded together before being screwed onto the telescope – a time-consuming process if your observing time is limited. Because the Sun's apparent diameter varies slightly during the year, due to the ellipticity of the Earth's orbit, the coronagraph comes with a set of six differently sized occulting discs. A different disc has to be threaded into the device every two months, to account for the Sun's changing diameter. Finally, to keep the Sun hidden behind the disc you need a driven equatorial mount that is precisely aligned with the pole. Accurate polar alignment is difficult during the daytime, when you cannot see the stars. This device is ideal for an observatory, where the telescope is permanently set up and polar aligned. But these disadvantages are out-weighed by the incredibly detailed views of prominences it can give. In particular, this device excels in photo-graphy. Unless you are prepared to pay very large sums, no other filter on the market enables prominences to be photographed in such detail with an ordinary camera.

Until the late 1990s, the only firm producing sub-angstrom H-alpha filters for the amateur market was the DayStar Filter Corporation, based in California, USA. DayStar filters worked the same way as Thousand Oaks and Lumicon filters, with an ERF with aperture stop at the front of the telescope and the main interference filter at the rear, ahead of the eyepiece. DayStar produced filters with a variety of passbands, from 0.8 down to 0.5 angstroms. All showed disc features as well as prominences, although those with the narrowest passbands gave the greatest contrast on disc features, with the disadvantage that prominences were fainter and so required longer exposures to be photographed using filters with the narrowest pass-bands. The very best (and most pricey) DayStar filters were electrically heated by a small oven controlled by a thermostat. The exact wavelength transmitted by a filter with a very narrow passband is strongly influenced by its temperature. The ability to control the filter's temperature meant that the DayStar systems gave very high-contrast images, but they were also inconvenient, as they required mains power. This was a particular nuisance for observers outside the USA, who needed a transformer to use the filter with their local voltage. However, DayStar also developed a line of sub-angstrom filters known as "T-scanners", which adjusted the wavelength transmitted by tilting the filter using a thumbscrew. Not only did T-scanners not require electricity, they were also cheaper. In the late 1990s, T-scanners cost from $1,300 to $2,300, compared to $2,250 to $5,600 for heated filter assemblies. ERFs cost $100 to $300 on top of this. Sadly, at the time of writing (late 2001), DayStar are no longer producing H-alpha filters. But many serious amateur solar observers are still using DayStar filters and so it is worth knowing how they work and how they compare to other filters currently available.

In the late 1990s another US firm, Coronado Instrument Group, began producing sub-angstrom H-alpha filters for amateur astronomers. Coronado filters (Figures 6.4 and 6.5) use a different design to DayStar. Both the ERF *and* the etalon are located in front of the objective lens, just like an aperture filter for white-light observing. In addition to the H-alpha line, all etalons transmit "side bands" of small parts of the spectrum at other wavelengths, and these have to be eliminated by a blocking filter behind the etalon. In the older filters the blocking filter was located directly behind the etalon, but in the filters produced by Coronado it is located at the eyepiece end of the telescope. Like the T-scanner systems made by DayStar, Coronado filters do not require electrical power and can be tuned precisely to the H-alpha line by tilting the blocking filter.

One advantage of Coronado filters is that because the etalon is located in front of the telescope, the f/ratio does not have to be increased to f/30, because the unfocused light rays are already parallel. Therefore the instrument does not have to be stopped down and a much higher resolution can be obtained. However, to obtain the maximum resolution possible in your telescope with a Coronado filter, the etalon needs to have the same diameter as your telescope's aperture. Therefore the larger your telescope the bigger – and more expensive – the filter needs to be. Coronado produce filters with apertures ranging from 40 mm to 140 mm, and their prices range from around £600 ($900) for the 40 mm version to £8,000 ($13,000) for their 140 mm aperture model! In practice, a Coronado filter may well stop down your aperture a little, as the filters are sold in a series of fixed apertures. Coronado offer a range of filters in between these price extremes, including the 60 mm aperture "Solar Max 60" ($2,000) and the 90 mm "Solar Max 90" ($4,000). The cheapest filter in the Coronado range, the 40 mm aperture "Solar Max", was introduced in the spring of 2001 and is a breakthrough in that it is the first sub-angstrom filter to be available for under £1,000 (or $1,000). It can be ordered to mount onto a number of small telescopes, including the Meade ETX and Celestron NexStar ranges. At a little extra cost, it can be ordered to suit *any* small telescope. While H-alpha filters are best suited to refractors, the smaller Coronado filters, including the Solar Max, can be used with Schmidt–Cassegrain or Maksutov telescopes. (Many amateurs successfully used DayStar filters with these types of telescopes too.) The Solar Max has a passband of 0.8 angstroms. Resolution with the 40 mm aperture is limited, but it is more than enough to reveal the principal disc features and prominences.

In addition to their line of filters, Coronado manufacture a complete H-alpha solar telescope, originally named the Helios I and later upgraded and renamed Nearstar. The design of this instrument is different to that of other Coronado systems in that the etalon is located inside the telescope, between the object glass and the eyepiece. The ERF is at the front of the telescope and has an aperture of 70 mm, the same as the object glass. A negative lens just in front of the etalon makes the light rays parallel before they enter it, and a positive lens behind the etalon (and blocking filter) reduces the effective f/ratio again to keep the tube short. The result is an extremely compact instrument with a focal length of only 400 mm – ideal if portability is important. Its price – £2,150 ($2,890) – is quite hefty, but its 70 mm aperture gives superb, high-resolution views of chromospheric features.

Figure 6.4. A Coronado ASP-60 (60 mm aperture) H-alpha filter mounted on a Tele Vue Pronto 70 mm aperture refractor. Photograph courtesy Dr David Boyd.

Figure 6.5. Close-up of the Coronado ASP-60 filter in its case, showing (left) the etalon to be mounted over the telescope aperture and (right) the blocking filter in a star diagonal to go in the eyepiece end of the telescope. Photograph courtesy Dr David Boyd.

What H-alpha filter you should buy depends on your interests in H-alpha work. An essential point to bear in mind with all H-alpha filters is that they are not mass-produced, like many amateur telescopes. Consequently, there are often waiting lists for these items and delivery times can be lengthy. The coronagraph I obtained from Baader Planetarium took five months to arrive, and a prominence filter I once obtained from another source took seven. If you are a mainly visual observer and want "a bit of everything" – i.e. prominences and disc features, a good choice of filter would be the Coronado model most appropriate for your telescope. An even better choice for visual observing would be a Coronado solar telescope (or a Coronado filter fitted to a small telescope) mounted "piggyback" on your main white-light telescope, thus allowing you to compare the view in white light with that in H-alpha simultaneously.

If your main interest is in photography, your requirements will be slightly different. Photography – and, especially, digital imaging – is possible with all the Coronado models, although the larger the filter the better, as a larger aperture gives higher resolution and shorter exposure times. However, if you want to obtain the best possible images of prominences, you may also wish to consider the Baader coronagraph. Prominences can, of course, be photographed through any sub-angstrom filter, but they are considerably fainter than the disc and so require a longer exposure, washing out the disc features and making the disc look distractingly bright. You could, of course, mask out the disc afterwards, either in the darkroom or by digital processing, but the resolution of the prominences will still be lower than with the coronagraph, because the longer exposure times cause details to be blurred by atmospheric turbulence. I will say more about H-alpha photography in Chapter 7.

Prominences and Filaments

Prominences can be divided into two broad categories – *quiescent* and *active*. Quiescent prominences are long-lived, lasting up to several months and surviving several solar rotations. They can be very large, running to hundreds of thousands of kilometres in length, although are usually only a few tens of thousands of kilometres high. The commonest type of quiescent prominence is known as a "hedgerow" and, as can be seen from the photograph in Figure 6.6, this is not a bad description, as they are long and low, connected to the Sun by a few short "trunks". They occur over the entire Sun, including very high latitudes where sunspots never

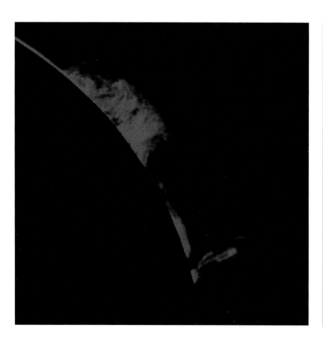

Figure 6.6. A hedgerow prominence, photographed by Derek Hatch using the same telescope and coronagraph as in Figure 6.2 but with a 2× and a 3× teleconverter stacked to magnify the prime focus image by 6×. Exposure $\frac{1}{8}$ second on Fuji Sensia 400 film in 35 mm format.

form. Over time the Sun's differential rotation tilts them so that they assume an east–west orientation and sometimes a number of such east–west tilted prominences join together to form a "polar crown", visible on the disc as a huge, long string of filaments across the Sun. Their numbers wax and wane in a cycle, but they are out of step with the sunspot cycle, their numbers being highest shortly after sunspot minimum. This is one feature that makes H-alpha observing so appealing: even at minimum, when there are few or no sunspots visible, there are usually some prominences.

Quiescent prominences are very stable, changing their appearance only slowly during their long lives, but they end their lives spectacularly when they lift off the Sun at speeds of up to several hundred kilometres per second and finally break up, all within the space of a few hours. When this happens, astronomers call the prominence an *eruptive prominence* or a *prominence eruption*. Astronomers still do not know what causes this type of prominence to suddenly erupt after several months of stability, but they believe that some eruptions are triggered by flares elsewhere on the Sun. Many amateur astronomers have witnessed prominence eruptions using H-alpha filters and they are very satisfying to record with a series of sketches or photographs made at intervals of, say, fifteen minutes.

A second type of prominence is associated with sunspots and appears as a loop above the spots. This is technically known as an *active region filament*, as when seen on the disc it appears as a filament winding through the sunspot group. Astronomers believe that such a filament marks the line separating opposite magnetic polarities in the group. Active region filaments can persist for days or weeks after the sunspots have disappeared and may evolve further to become quiescent prominences.

Active prominences are associated with solar flares and can change their appearance over periods of a few minutes. A common type is a *surge*, which appears as a short-lived "spike" at the solar limb. This involves material shooting up into the (invisible) inner corona in a straight path at speeds similar to those in erupting quiescent prominences and then descending back into the chromosphere in the same path. The fastest-changing prominences of all are known as *sprays*, which indeed resemble a spray of glowing red hydrogen, reminiscent of a volcanic eruption. The contents of a spray prominence can move outwards faster than the Sun's escape velocity and so are ejected from the Sun altogether. Flares can also produce small, brilliant loop prominences, composed of material descending back into the chromosphere. Note that despite their spectacular eruption speeds, prominences do not generally change their appearance before your very eye, due to the Sun's great distance from us. It usually takes a minute or two for changes to become apparent in a telescope.

I have noted above that the chromosphere appears as a thin band of light around the limb of the Sun. However, if you examine it using an H-alpha setup with reasonably good resolution you may notice that its outer edge appears somewhat furry. It is, in fact, full of tiny, prominence-like spikes, known to astronomers as *spicules*. These behave like prominences in that they are composed of material rising in columns and then gradually fading from view, but they are part of the general structure of the chromosphere and form all over the Sun. They are not associated with sunspot groups or active regions and so are not counted as prominences in prominence statistics.

The simplest form of useful observation you can make of prominences is to count the number of prominences you see each day and work out a prominence MDF at the end of each month, just as you would do for sunspot groups in white light. The exact method of counting varies with different observing organisations. The BAA Solar Section method is to divide the limb into latitude zones of 5°, each zone counting as one prominence active area if it contains one or more prominences. It can, however, be difficult to establish what is 5° in your eyepiece, unless you have a reticule or some other way of making positional measurements (see below). Whatever method you choose, prominences do tend to occur a good deal further apart than this, whether they form singly or in clusters of several prominences, and it is usually quite easy to tell whether or not a prominence counts as an active area. Occasionally you will see prominences that appear to be detached altogether from the chromosphere, and these do not count as active areas.

Drawings and position measurements for prominences are much harder to do than for sunspots. In H-alpha, we do not have the convenience of projecting the Sun's image onto a grid and copying the positions of features down. The images produced by an H-alpha filter are much too faint to project and so we are constrained to drawing features as seen through an eyepiece. One method is to draw the prominences after you have drawn the sunspots using the white-light projection method. You can then use the positions of the spots as a reference frame on which you can plot prominence positions. But accuracy is always a compromise with this method, especially if there are not many sunspots present. If you use a coronagraph, this method would be very difficult, as the Sun's disc is then masked. An ideal measuring instrument is a reticule, either silhouetted against the Sun's disc or an illuminated reticule seen against the black sky surrounding the Sun, with angles marked out at intervals of 5° or less. Using such a reticule you could draw prominences in the same way as you would draw sunspots in white light. First, establish the orientation of the image using the drift method. Then, using a protractor, draw a pattern identical to that of the reticule around the edge of the grid you use for solar drawing, and overlay a sheet of thin paper as before. You can now copy down the positions of the prominences seen in the eyepiece as easily – and as accurately – as you would for sunspots on a projected image. The trouble is that a suitable reticule can be hard to come by. One excellent device is an eyepiece designed for guiding long exposures of deep-sky objects. This has an illuminated reticule which includes, among other things, a 360° circle around the circumference of the field, marked off at 5° intervals – ideal for prominence drawing. Unfortunately, the main purpose of these eyepieces means that they give a fairly high magnification, their focal lengths usually being 12.5 mm or shorter. This is no problem on a small telescope with a short focal length; for example, the Coronado Nearstar telescope gives a magnification of only 32× with a 12.5 mm eyepiece. But in instruments with longer focal lengths the magnification will be too high to show the whole Sun in the field of view – essential for measuring prominence positions. Another possibility is the pole-finding eyepieces supplied with some up-market finder-scopes, as these tend to have lower magnifications. You could also try your luck at government surplus optical suppliers, which sometimes supply reticule eyepieces originally intended for use with microscopes or military optics. Other amateurs have invented various ingenious ways of measuring prominence positions indirectly and it is also possible to make measurements from photographs or digital images.

A somewhat easier way of recording prominences is to make a detailed drawing of a particularly interesting prominence (Figures 6.7 and 6.8). Because we are not measuring their positions relative to the whole Sun, precise measurements with a reticule or other means are not required. Establishing the directions of north and east using the drift method is all that is necessary. We can use relatively high magnifications here, because we are not viewing the whole Sun. While drawing by hand is not as accurate as photography or CCD imaging, it is often more convenient, as setting up the camera and focusing the image can be very time-consuming, particularly if clouds are threatening to roll in and end observations. Making an accurate prominence drawing is not easy, however, because when viewed through a good instrument prominences can be very intricate. It is best to start by drawing relatively small and simple prominences, before progressing to more complex types. To ensure consistency and ready comparison with other drawings, try to do all your prominence drawings to the same scale. Represent the Sun's limb by drawing a segment of a circle of known diameter. To begin with, you could use a 6 inch protractor for this and draw prominences to the same scale as full-disc drawings in white light. Begin a drawing by outlining the basic shape of the prominence as accurately as possible and then fill in the finer details.

It is possible to estimate the heights of prominences, using drawings or photographs. Simply measure the height of the prominence on your image in millimetres or inches and then divide it by the diameter of the disc you are using. Multiplying the result by the Sun's true diameter (1,392,000 kilometres or 864,000 miles) gives the actual height of the prominence. A simple measurement such as this reminds us of the vast scale of solar features. Even quite small prominences dwarf the Earth, whose diameter is 12,756 kilometres.

If you have a filter capable of revealing disc features you can record filaments visually by drawing them on a white-light sunspot drawing form, after you have made a sunspot drawing and so using the spots as a guide to positioning the filaments. It is interesting to distinguish different types of filaments and their equivalent appearance as prominences on the limb. Quiescent filaments tend to be long and thick, although they are sometimes fainter than some other types. When they are near the limb, filaments can appear three-dimensional and quiescent filaments sometimes show

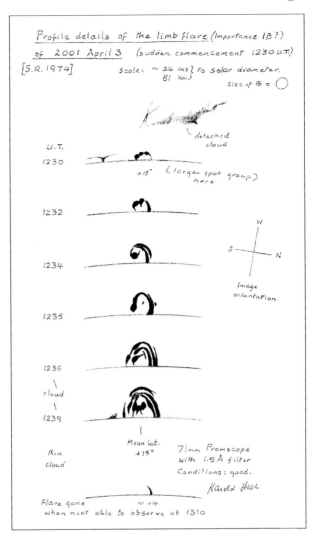

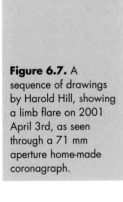

Figure 6.7. A sequence of drawings by Harold Hill, showing a limb flare on 2001 April 3rd, as seen through a 71 mm aperture home-made coronagraph.

the "hedgerow" appearance of quiescent prominences. Filaments associated with sunspot groups are thin and dark, often with a winding, snake-like appearance, and they sometimes show up as loop prominences at the limb. Note also how the chromospheric structure changes close to sunspot groups (Figure 6.9). The granular structure is distorted and it often shows the familiar pattern formed by iron filings over a magnet.

One exciting aspect of observing filaments is watching them rise off the surface of the Sun. Like prominences, filaments move too slowly for their motion to be seen directly, but you can tell they are

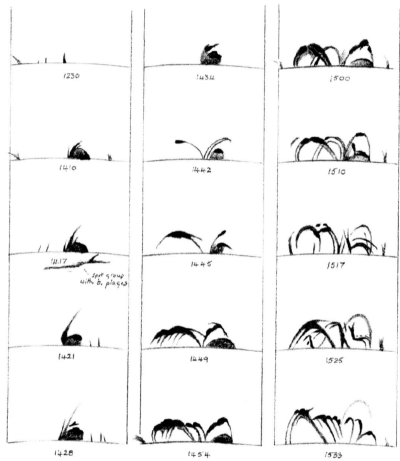

Development of a loop system, following a flare (1410 u.t.) at the SW limb over the naked-eye spot group at mean lat. -21° long. ~0° nearing limb passage on <u>2001 April 15</u>.   Synodic rotation 1975.
(see loop system over same area at E. limb on Apr. 1 & 3)

1230   1434   1500

1410   1442   1510

1417   1445   1517
spot group
with b. plages

1421   1449   1525

1428   1454   1533

Positions derived by peripheral graticule on X90 wide-angle ocular.   Scale: 24 in/81 km to solar dia.

A lot of cloud now from the north.. End of record.

71 mm Promscope; coronagraph-type; 1.5Å filter & drive.   Harold Hill

Figure 6.8. A sequence of drawings by Harold Hill of an active prominence system on 2001 April 15th. Equipment the same as in Figure 6.7.

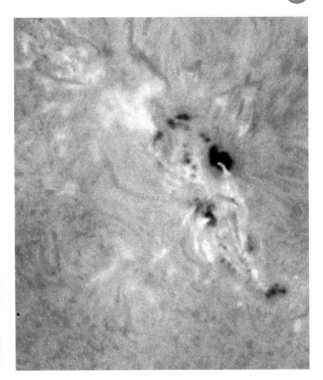

Figure 6.9. The great sunspot group of March 2001 – the same group as shown in Figures 3.4 and 8.1 – as seen in H–alpha. The granular structure of the chromosphere is distorted near the spots and numerous plages are visible. CCD image by Eric Strach, using a 200 mm SCT with DayStar filter as in Figure 6.1 and a Starlight Xpress SXL8 CCD camera.

moving using the Doppler effect – the compression or lengthening of the wavelength of light emitted by an object moving towards or receding from us. In astronomy, the Doppler effect is best known for its effect on the light of distant galaxies receding from us at very high velocity. As the wavelength of light emitted by the receding galaxies lengthens, the light becomes more and more red, because red light is simply light with a longer wavelength. Thus the effect is known as "redshift". Conversely, light emitted by objects moving towards us is *blue*-shifted, as its wavelength is reduced. We can observe blue-shifting on a smaller scale on the Sun. When an H-alpha filter is tilted slightly away from the focal plane, its peak transmission is shifted away from the H-alpha line, towards the blue end of the spectrum. If a filament is rising towards us, it appears fainter when the filter is aligned square-on to the optical axis but becomes more prominent when the filter is tilted "off band". The effect is most dramatic when a prominence erupts while it is visible as a filament on the disc. The filament then rapidly disappears – not because it has broken up and its contents dispersed, but because it is moving so fast that it has been blue-shifted into the violet part of the spectrum, way outside the range of an H-alpha filter. Such an eruption when seen on the disc is sometimes known by the French term *disparition brusque* – literally, a "sudden disappearance". Doppler shifting is most noticeable in filaments near the centre of the disc, which are moving directly towards us.

Flares

Flares usually occur in or near sunspot groups, but they are occasionally seen in active regions before any sunspots have formed or after they have all disappeared. In Chapter 3 we noted that the extremely rare white-light flares are most likely to occur in complex groups, of types D, E and F in the McIntosh system, and the same is true for flares in H-alpha. Very large and complex groups, such as that of March 2001 shown in Figure 3.4 (Chapter 3), can produce many flares during the course of their two-week transit across the Sun and if such a group is present it is worth checking it for flares with your H-alpha filter several times a day, if you can.

In both their frequency and their intensity solar flares broadly follow the rise and fall of the solar cycle, although professional astronomers have sometimes observed a slight resurgence in flare activity in the late stages of the cycle, when sunspot numbers are dropping. At solar maximum, several hundred flares can occur each month, although you will not record anywhere near this number with your telescope, because flares tend only to last for a few minutes and many will take place when the Sun is not visible. At minimum, only a small handful of flares occur in a month and days or even weeks can go by without a single flare being recorded. Also, flares at this time are generally much smaller than those observed at maximum.

A flare often begins as a pair of bright points of light in or near a sunspot group. These then brighten rapidly and increase in size, often elongating into short streaks or "ribbons", before gradually fading from view. Flares generally take much longer to fade out than they do to reach maximum brightness. The majority of flares have lifetimes measured in minutes, but very large and active ones can last for several hours. Remember not to confuse the bright plages around sunspots with flares;

these tend to cover a much larger area and change much more slowly.

When you see a flare, it is important to record several features. Because flares are transient events, it is essential to time their appearance accurately, to the nearest minute. Note down the time you first noticed the brightening, the time at which it reached its maximum intensity and the time at which it had faded from view and you could no longer see it. As with all solar observations, make a note of the observing conditions and the telescope and H-alpha filter you are using.

Also important is an estimate of the flare's position. The most practical way of doing this is to sketch the flare's appearance relative to its parent sunspot group. A photograph or CCD image is, of course, more accurate, but setting up the camera may take too much time in a short-lived event such as this. When you have made a sketch of the flare it should then be quite easy to get an accurate estimate of its latitude and longitude from a disc drawing showing the group's position on the Sun.

Astronomers use a system for classifying the appearance of flares in H-alpha light, based on the flare's peak intensity and the area of the Sun's disc it covers. The flare's intensity relative to its surroundings is assigned using one of three letters: *f* (faint), *n* (normal) and *b* (bright). A brightening not much greater than the surrounding chromosphere would be described as *f*, while *b* refers to something strikingly brilliant. The area in square degrees covered by the flare when at its peak intensity, sometimes known as the flare's "importance", is described by the numbers 1–4, as in Table 6.1.

Flares covering an area of less than 2 square degrees are known as *sub-flares* and are denoted by the letter S.

When a flare is classified, the number representing its importance is written before its intensity. For example, a large and bright flare covering 15 square degrees would be class 3*b*. The flare photographed by Eric Strach in Figure 6.10 is of class 2*b*. In practice, it is

Table 6.1

Importance	Area covered (square degrees)
1	2–5
2	5– 2.5
3	12.5–25
4	over 25

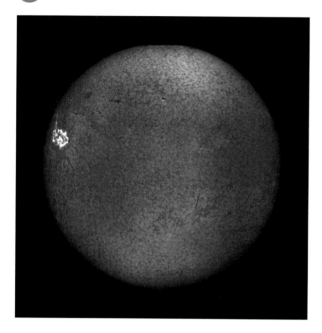

Figure 6.10. A class 2b flare photographed by Eric Strach on 1999 July 19th. Exposure 1/90 second on TP 2415 film, using a 200 mm SCT and 0.6 angstrom passband DayStar filter. The Sun's disc has been deliberately underexposed at the printing stage to bring out the brilliance of the flare.

difficult for a visual observer to make an accurate estimate of a flare's intensity and especially its area. It is best to make sketches as described above and leave the final classification to the director of your observing group. You can, however, make a rough estimate of a flare's class using your own sketches and its size relative to the disc drawing.

Observing the Sun in H-alpha is fascinating – and, indeed, is a whole subject in itself. But don't let it become all-absorbing and replace white-light observing, as the latter is the main "bread and butter" of amateur observing groups and is the area in which they need the most observations. My own rule is always to do my white-light observing first and only when this is completed do I go on to H-alpha if time and weather permit. But the views to be had in H-alpha are so spectacular that it is often tempting to break this rule, especially when activity is high!

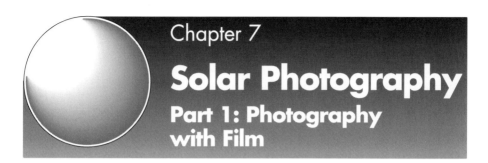

Chapter 7

Solar Photography
Part 1: Photography with Film

When it is done correctly, solar photography can be very rewarding. It is the most accurate way of recording sunspots and other solar features – although, as I have explained in Chapter 4, it is very difficult to do accurate positional measurements using photographs. Nevertheless, photographs of the Sun's full disc are excellent for showing what was visible on the Sun on a particular day and show the appearance of the Sun's features far more realistically than drawings ever can. Photographs shot on slide film are great for demonstrating recent solar activity to astronomy clubs. Photography is the only way of getting a pictorial record of the Sun if there is too little time to make a drawing – for example, if clouds are approaching. Whereas a disc drawing can take half an hour or more to complete if the Sun is active, the Sun can be recorded with an exposure in the region of $\frac{1}{250}$ second! Of course, mounting the camera and focusing the image takes up some time, but even allowing for this, photography is often far quicker than drawing. Especially useful are close-up photographs of individual sunspot groups, as drawings showing the details of complex groups are not easy to do accurately. And if you have an H-alpha filter, it is possible to take dramatic shots of prominences and other chromospheric features.

One of the beauties of solar photography is that, as with other forms of solar observing, superb results can be obtained with just a small telescope (Figure 7.1). Also, unlike most kinds of night-sky photography, exposures for the Sun are so short that your telescope doesn't need to be motor-driven to counteract the Earth's rotation. I did many successful early experiments in solar photography using just a basic 60 mm alt-azimuth mounted refractor costing less than £150. Similarly, the type of camera you use does not have to be sophisticated or expensive. An ordinary, good-quality, 35 mm single-lens reflex (SLR) or digital camera from your local camera store is all that is needed.

Until the 1990s film was the only medium with which amateur astronomers photographed the Sun. But in recent years digital photography has assumed an increasingly important role. You can photograph the Sun with two types of digital camera – the ordinary digital camera as found in camera shops, intended for everyday photography, and the more specialised astronomical CCD camera, designed for astronomical imaging through telescopes. From the viewpoint of the amateur solar photographer, digital imaging has many attractions. Chief among them is the fact that the results are instant: you don't need to wait for the pictures to be developed and you don't need a darkroom to develop the images yourself. But traditional film photography has a number of advantages over digital. Conventional cameras do not require the use of a computer to store and process the images. Secondly, all digital cameras use up considerable quantities of battery power, whereas many of the best conventional cameras for solar photography are fully mechanical and require no

Figure 7.1. An example of what can be achieved with a small telescope. Three sunspot groups, photographed by the author on 1993 May 28th, using a 60 mm f/11.7 refractor with eyepiece projection, giving an effective focal length of 5,800 mm and an effective f/ratio of f/102. Exposure 1/250 second on Kodak Ektachrome 400HC film.

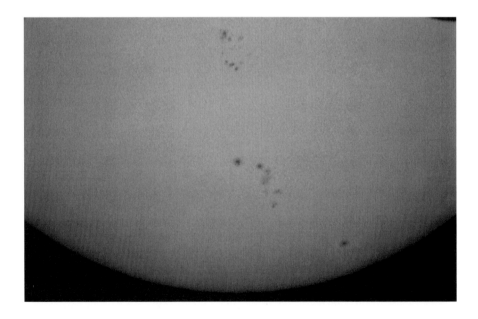

power at all. Thirdly, 35 mm film has better resolution than the CCD chips inside most digital cameras. Only very expensive models intended for professional use have chips even approaching the resolving power of 35 mm film, and the resulting images require enormous computer power to process. Fourthly, most amateur-level digital cameras do not have removable lenses, which makes many models difficult to mount on telescopes and restricts the range of photographic techniques it is possible to use. Finally, a conventional camera is good value for money in that it is equally suitable for *all* kinds of photography – lunar, planetary, deep sky and everyday photography. While they are very suitable for solar work, most digital cameras are of more restricted use in other kinds of astrophotography. Apart from the Moon and the brighter planets, most celestial objects are too faint to be imaged with commercial digital cameras. An astronomical CCD camera is, of course, ideal for night sky imaging, but is much less suitable for everyday shooting or wide-angle astronomical work.

Because film and digital photography both have their strengths and weaknesses, I have divided this chapter into two parts. In this part I will deal solely with emulsion photography and in the book's final chapter I will explore the possibilities that digital techniques have to offer the solar observer.

Equipment

Solar photography can be almost as simple as everyday photography. If you use the projection method to observe the Sun, you can photograph the projected image quite easily. As the projected image is brighter than its surroundings, you will get the best results if you use a slightly shorter exposure or slightly smaller f/stop than you would for normal photography in sunlit conditions. Shooting the projected image has two major problems, however. The first is that you can never photograph the Sun's image from directly above, since then the eyepiece would be in the way. Therefore the solar disk will be somewhat distorted in the photograph. Secondly, it is difficult to get good contrast in photographs of a projected image, and sunspot detail often appears washed out. The resulting photograph also tends to be rather "messy", with telescope parts and other objects intruding into the picture. You can correct the distortion by placing the camera as close as possible to the telescope drawtube and tilting the projection screen towards the camera. Tilting, however, will throw parts of the solar image slightly out of focus. An alternative could be to scan the image and use digital processing techniques to eliminate the distortion. You can also increase the contrast by projecting the image onto a blank wall or perhaps a slide projection screen in a darkened room. However, no photograph of the projected image I have ever seen even comes close in quality to an image shot directly through a telescope. Photographing projected images at high magnification is especially difficult, as the contrast is then even lower. The technique is useful in emergencies, though – for example, when a projected image showing a large sunspot is on display at an astronomy club gathering and you are unable to photograph the Sun with your own telescope.

Serious solar photography needs to be done directly through the telescope. A word of caution here: **solar photography through the telescope requires exactly the same safety precautions as for visual observing – i.e. aperture filters designed for solar observing with telescopes**. Without the proper filters, the Sun is every bit as dangerous when seen in a camera's viewfinder as it is when viewed with an eyepiece!

Filters

Your choice of aperture filter for photography is important, as some filters give much better results than others. Most filters for visual use have a "neutral density" of 5, which means that they transmit 1/100,000th, or 10^{-5} of the Sun's light. This usually gives an image that is bright enough for visual observing, even with quite high magnifications, but it can often be too dark for photography, particularly when you are shooting a highly magnified solar image to capture details of sunspot groups. A faint image needs a longer exposure to be recorded. But long exposures leave us prone to a major problem affecting all kinds of astrophotography in which a camera is attached to a telescope: vibration. When the shutter in an SLR camera is fired, two things happen in quick succession: the shutter opens and the mirror reflecting the image into the viewfinder flicks upwards, out of the light path.

Both actions, especially the mirror movement, cause vibrations. These are too small to be noticed when the camera is used for everyday photography with normal lenses, but at the high magnifications used in telescopes even tiny vibrations are enough to blur the image. Although some cameras have a much smoother shutter release and reflex action than others (as I will explain below), all cameras produce some vibration when used with a telescope. There are two ways of getting round this problem. One technique, well-known among lunar and planetary photographers, is the "hat-trick", in which the telescope aperture is covered by a black card held over the front (but not touching it) and the shutter released while the telescope is covered. A few seconds later, when any vibrations caused by the camera have died away, the card is removed. At the end of the exposure the card is replaced and the camera shutter closed after the telescope is covered. The trouble with this method is that human reflexes can only accurately time an exposure of about half a second or more, whereas solar exposures are generally much shorter than this. You could increase the required exposure time by placing a second filter in the system, such as a dark photographic neutral density filter *in addition to* the aperture filter. But long exposures such as this are usually spoiled by the daytime seeing conditions – which, as explained in Chapter 3, can often be poor. There is also the problem of daylight reflecting off the card covering the telescope.

The second, and much better, method, is to use an exposure too short for any vibrations to register. Generally, exposures of $\frac{1}{125}$ second or less come into this category, although sometimes you can get away with $\frac{1}{60}$ second or even longer with the correct camera and telescope setup. Such short exposures have the additional advantage that the effects of bad seeing conditions are greatly reduced. For shooting the whole disc, exposures of $\frac{1}{125}$ or $\frac{1}{250}$ second are possible with many visual filters and a medium-speed film. But the exposure required can often be longer. Between $\frac{1}{125}$ second and $\frac{1}{2}$ second exposures are long enough to cause vibrations but too short for the hat trick to be of any use. Many solar exposures using visual filters at high magnifications fall smack in this range. You could use a faster film to reduce the exposure time, but this greatly increases the film grain and in solar photography we want to capture as much fine detail as possible. The solution to the problem is to use a filter

that transmits more light. I have found that the Baader
AstroSolar filter, in addition to being a first-rate visual
filter, also gives a somewhat brighter image than other
visual filters. It is excellent for whole-disc photography
and can even be used at higher magnifications showing
only part of the disc. If you stick to low to medium
magnifications, a visual filter giving a reasonably bright
image may be all you need. Even better is a filter
specially designed for photography. Baader make such a
filter, metal-coated on polymer like their visual filters,
called AstroSolar Photo Film ND3.8, and the Type 3+
filter produced by the US firm Thousand Oaks Optical
is an example of a metal-on-glass photographic filter.
Both filters are coated to a neutral density of
approximately 4 – i.e. they transmit 10^{-4} of the Sun's
light, ten times more than visual filters. But these filters
are *not* safe for visual use, even for focusing the image
through the camera viewfinder. Always focus the image
using a visual filter first, and then switch to the
photographic filter when you are ready to make the
exposure. With the correct filter, exposures of $\frac{1}{250}$, $\frac{1}{500}$ or
even $\frac{1}{1000}$ second are possible, even at higher magnifications. The prices of photographic filters are approximately the same per aperture as for visual ones. One
small problem with the Baader photographic filter is
that it is not available in A4-sized sheets, like its visual
cousin, but comes in a roll 1 metre long and 0.5 metre
wide, costing over £50. One solution to this could be to
buy a roll collectively with other members of your local
astronomy club.

Telescopes and Mounts

You can use any of the three main telescope types for
solar photography. In theory, a refractor should have
the edge, as it has no central obstruction and so the
image contrast should be higher, but in practice all
telescopes can give excellent results. Because you are
using an aperture filter, there are no problems with heat
build-up in Newtonians, Schmidt–Cassegrains and
Maksutovs. If you use a Newtonian, however, you
may find that you cannot rack the focuser in far enough
for the image to reach focus in your camera's
viewfinder. One remedy for this is to substitute your
existing focuser with a "low-profile" device intended
for astrophotography, which enables the camera to be
placed closer to the tube. Alternatively, use a Barlow

lens between the camera and the focuser. This, of course, increases the size of the image and the effective f/ratio, and therefore the exposure, but as many modern Newtonians are relatively "fast" (f/8 or faster), this is not usually a problem.

Whatever type of telescope you use, if it is larger than 150 mm (6 inches) you need to stop down the aperture. At most observing sites, the seeing conditions during the day rarely permit resolution higher than the theoretical resolving power of a 100 mm (4 inch) telescope – i.e. about 1 second of arc. Solar images taken at full aperture using a large instrument tend to be blurred and have a washed-out appearance. Stopping down will improve the contrast and sharpness of the image, even though the theoretical resolution will be lower. If you have a Mylar-type filter, you can stop down the aperture by making an off-axis mask from cardboard as described in Chapter 2. For photography, though, you need to use two layers of cardboard, each with a hole of the same size, with the filter material sandwiched between them. Glass filters usually come ready-made, with the filter already mounted in the mask, which is sized to suit a specific aperture and make of telescope.

More important than the telescope type is its mount. If you are to reduce image-blurring vibrations to an absolute minimum, you must use as steady a mount as possible. This is yet another reason for avoiding very cheap telescopes, as many of them are sold on flimsy mountings. Fortunately, many "serious" astronomical telescopes are nowadays sold on mounts designed with astrophotography in mind and the requirements for solar photography are not as stringent as those for night-sky work, which demands much longer exposures. My own 80 mm refractor is mounted on Celestron's CG-4 equatorial head, a Chinese equivalent of Vixen's Great Polaris mount, and I have found this to be quite adequate for the job. It helps to ensure that there are no loose joints in the mount – for example, by tightening the thumbscrews connecting the tripod to the equatorial head.

As I have said above, a motor drive is not essential for solar photography. To avoid vibration, you need to stick to exposures of $\frac{1}{125}$ second or faster and these are too short for the image to be blurred by the Earth's rotation, even at very long focal lengths. It has to be said, however, that a drive is very convenient, as when you are setting up and focusing the camera the Sun quickly disappears from view if the telescope is not driven. Electronic slow motion controls also make it easier to centre individual sunspot groups at high magnification.

Cameras

As with other areas of astrophotography, much the best type of camera for solar work is the 35 mm single-lens reflex (SLR). This type of camera allows the image to be viewed exactly as it will appear in the photograph up to the moment you fire the shutter and a huge range of films and accessories are available for it. 35 mm SLRs can also be attached to telescopes easily, because their lenses are removable. They permit a wide range of shutter speeds, from the "B" setting (for time exposures longer than one second) right down to $\frac{1}{1000}$ second and sometimes even faster. Unfortunately, simple cameras, such as the popular 35 mm compact variety, are quite unsuitable for solar work, because you cannot remove the lens or control the exposure adequately. "Medium-format" cameras, such as the Hasselblad and Mamiya makes, are prized by some advanced photographers, because the larger format of the film allows more of the Sun's disc to be photographed without sacrificing resolution. However, they are bulky on small telescopes and can be very expensive.

Most SLR cameras can be used to some extent for solar photography, but some are much better than others. You do not need one of the latest electronic models, with automatic focusing and exposure, as these features will not work for solar photography through a telescope. That is not to say, however, that your camera has to be fully mechanical, as is recommended in many astronomy books and magazine articles. This advice is very sound for night-sky photography, because keeping the shutter open for a long exposure quickly drains the batteries, especially in the low temperatures often encountered at night. But in solar work we are using very short exposures, mostly in relatively warm conditions. One feature requiring battery power, motorised film advance, can actually be an advantage for solar work, since it allows numerous exposures to be taken in quick succession and could be useful for time-lapse photography in H-alpha. That said, it is always a good idea to have a fully mechanical camera at least as a standby, as there is much less to go wrong in these models. It is very frustrating to have to terminate a photographic session due to a flat battery or some malfunction in the camera's electronics.

Ideally, an SLR for photographing the Sun should have three important features. Most important of all is the ability to take very short exposures, from $\frac{1}{125}$ second down to $\frac{1}{1000}$ second and even shorter if possible. Also important is low vibration when the shutter is released. In addition to a smooth shutter, to reduce vibration to a minimum it is desirable for a camera to have "mirror lock", in which the mirror can be flipped up manually with a lever well before the shutter is fired. Only some cameras have this feature. On some other models, the self-timer feature flips up the mirror at the same time as you press the shutter release, a good few seconds before the shutter actually opens, allowing time for vibrations to die down. A third desirable feature is the ability to change the focusing screen underneath the camera's prism. On most cameras, the central part of the screen, on which you make critical focusing adjustments, has a coarse texture. This is fine for everyday photography, but a telescopic image can be too faint to focus well on such a screen. The split-image focusing facility found on some other cameras is even worse. Fortunately, some camera manufacturers offer a range of focusing screens for different applications. A good choice for solar photography is a screen with a clear central section on which is engraved a fine cross, allowing your eye to focus at the correct point. Some manufacturers of astronomical accessories sell special bright screens for astrophotography. Even if your camera does not allow the screen to be changed, a few companies advertising in astronomy magazines offer a screen polishing service, in which a central clear spot is made.

Many good cameras for solar photography are no longer manufactured but are often available as used cameras in quality photographic stores. When buying a camera, try to get a knowledgeable member of the shop's staff or an experienced amateur astrophotographer to check whether it has the required features. A good second-hand camera can be quite cheap. For example, my Cosina CT-1 cost less than £100 and has served me well for many years. Its focusing screen is coarse-ground, which makes focusing difficult at high power, but it has exposures down to $\frac{1}{1000}$ second and the self-timer releases the mirror at the beginning of the self-timer interval. My favourite camera, however, is a second-hand Olympus OM-1. Long out of production, this particular camera is coveted by many astrophotographers, because it is fully mechanical, with an exceptionally smooth shutter release and mirror movement, light weight and the ability to accept a large range of readily available accessories, including different focusing screens. It also has mirror lock, although this feature is fiddly to operate on the OM-1. In any case, the mirror movement on this camera is so smooth it doesn't need it. The later models in the OM series, such as the OM2, OM3Ti and OM4Ti, have many of the same features. They are battery-operated, but this is of no consequence in solar photography. Other cameras with features useful in solar photography include the Nikon F and FM lines. Later Nikon F models have exposures shorter than 1/1000-second, which could be useful for photographing a very bright solar image.

I will mention some accessories for cameras when I come to discussing the various techniques for photographing the Sun below. One accessory you cannot do without, however, is a cable release. It is impossible to press the shutter release button directly on any camera attached to a telescope without causing serious camera shake. Using a cable release ensures that your hand never touches the camera when making the exposure. Cable releases are available cheaply in any good camera store. Try to obtain the longest one you can find. A 300 mm (12 inch) is satisfactory, but a 500 mm (20 inch) is even better.

Films

The Sun can just as well be photographed in black and white as in colour, because the Sun is essentially a black and white object and aperture filters tend to give the image a false colour cast. Indeed, black and white films have an advantage over colour in that they have finer grain, and therefore higher resolution, than colour films of the same sensitivity. Much the best black and white emulsion for solar work is Kodak's Technical Pan 2415 film. Technical Pan was originally designed for scientific applications and it combines extremely fine grain with high contrast. The latter is especially useful for recording sunspots, which have a fairly low contrast (Figure 7.2). The trouble with black and white film is that it is best to process it yourself. These days most commercial and even professional processing laboratories are geared towards colour films, and many laboratories simply do not do black and white. If you are lucky enough to find a suitable black and white service, you will soon find that getting them to produce prints on a regular basis is expensive. The commercial black and white service offered by some high street camera stores is satisfactory for processing negatives, but the prints, which are produced by a mechanised process, are of very poor quality. To produce black and white prints at home you need a darkroom equipped with an enlarger – something that not all of us have the money or the space for. If you are fortunate enough to have such facilities, then for many purposes black and white is the way to go. I will not discuss developing and printing in detail here, as it is amply covered in books on general astrophotography, as listed in Appendix C and, of course, in everyday photography books.

Because of the difficulty of processing black and white, many solar photographers are attracted to colour films. Avoid colour print film, if at all possible. This type of film gives low-contrast images, making it unsuitable for recording solar features. Moreover, it is difficult to get decent colour prints made commercially. Most processing laboratories are not used to processing astronomical negatives, let alone shots of the Sun, and many commercial services make prints using an automated process. If a film contains nothing recogni-

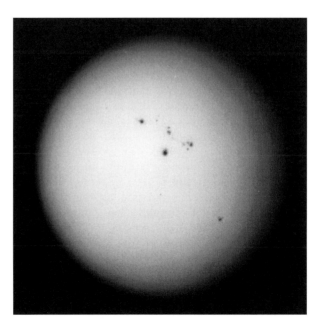

Figure 7.2. Whole-disc photograph, showing four spot groups spaced close together, taken by the author on 2001 June 19th, using an 80 mm refractor, Baader AstroSolar Safety Film filter at full aperture, and eyepiece projection giving f/33 (effective focal length approx. 2,600 mm). Exposure 1/125 second on Kodak TP2415 film. Note how the film's high contrast exaggerates the limb darkening.

sable, you may well end up with no prints at all. Even worse, because solar shots often have a black background, the technician or machine might not be able to see the boundaries between the frames on the negatives and so cut through some important photographs. It is always best to use slide film. Developing slides is a one-stage process and so there is no need to worry about prints. To avoid having any images cut through, you can ask for the film to be returned uncut and then mount the slides yourself at home. Even better, slide films have high contrast and, although they are still outclassed by the best black and white films, today's low-sensitivity slide films have very fine grain. Slides have several other advantages. They are great for illustrating astronomy club talks and lectures to the general public. You can make duplicates quite easily using a slide copier attachment to your camera. Slides are also easy to store and you can view them with inexpensive slide viewers.

To obtain images with the finest grain possible you should use slow slide films with ISO values between 50 and 200. Try to avoid using faster films, as they have more noticeable grain. It is better to use a brighter filter instead, if possible. A 400 ISO film is, however, acceptable if you are shooting at high magnification with a very small telescope. I took the picture in Figure 7.1 with a 60 mm refractor on Ektachrome 400 (now known as Elite Chrome 400) film. It is always better to have a slightly grainy but sharp picture than one blurred by camera vibration or atmospheric turbulence. My preferred films are Elite Chrome 100 and 200. You could also try the Fujichrome Provia 100 and 200 ISO films, or the very fine-grained Velvia 50 ISO. But avoid process-paid films such as Fujichrome Sensia or the Kodachrome range, as to take advantage of the process-paid offer you need to send the films away through the post and so risk losing original photographs. I once had to wait an agonising three weeks for a roll of process-paid film to be returned. Similarly, avoid using any processing service that sends films away to a central processing laboratory. It is far better to use a local professional laboratory that develops films on the spot. Not only does this ensure that your slides are not lost, but the waiting time is also much shorter. My local laboratory will process a slide film and have it available for collection the same day. Also, there is only a slight difference in price between professional laboratories and high street services.

Photographic Techniques

Prime Focus

The simplest method of photographing the Sun is known as the "prime focus" method, in which the eyepiece is removed and the camera attached directly to the telescope, using the telescope as a long telephoto lens. It is, in fact, possible to get reasonable images of the Sun using a very long telephoto lens system intended for photography, say a 500 mm telephoto with a 2× or 3× teleconverter – always provided that the front of the lens is covered with a safe aperture filter. Such a setup can provide shots showing the Sun's full disc with the larger sunspots. But to take detailed sunspot photographs you need a telescope.

When photographing the Sun (or Moon) astrophotographers use a rule of thumb, which says that the size of the Sun's disc on the film, in millimetres, is equal to the focal length of the lens used divided by 110. The 35 mm film frame measures 24 mm × 36 mm, so for the Sun to fill a good part of the picture and show details in the sunspots and other features you should aim for a focal length of at least 1,000 mm and preferably rather more. In prime focus photography, the focal length is simply the focal length of the telescope, which for commercially made telescopes should be printed both on the body of the instrument and in the instruction manual.

Prime focus photography gives excellent whole-disc shots showing the appearance of the Sun on a particular day. For example, with a Meade ETX 90 mm Maksutov telescope (focal length 1,250 mm), shooting at prime focus gives a solar disc about 11.5 mm across, capable of showing plenty of detail on fine-grained film. A "classic" 150 mm (6 inch) f/8 Newtonian reflector gives an image of about the same size. A 200 mm (8 inch) f/10 SCT, of 2,000 mm focal length, is even better, producing an 18 mm disc. Only in telescopes with focal lengths longer than 2,500 mm, such as large SCTs, is the Sun's disc too large to fit in the 35 mm frame – although the largest manufacturers of these instruments, Meade and Celestron, offer devices to reduce the telescope's focal length and so in many cases allow these instruments to photograph the whole disc.

To take prime focus shots, you need to couple the camera to the telescope using a camera adapter designed for this purpose. The adapter itself attaches to the camera using a "T-ring", a device with a bayonet fitting to slot into the camera like a lens. The front of the T-ring has a standard thread which should be the same as the thread on the adapter. As with camera lenses, T-rings vary and you need to get the one that fits your camera. T-rings for a number of different makes of camera are available from a good camera store, while you can buy camera adapters from many of the equipment suppliers advertising in astronomy magazines. The type of adapter you need, however, depends on the type of telescope you are using. If you have a refractor or a Newtonian, you need a standard camera adapter, the front of which pushes into the drawtube like an eyepiece. This type of adapter unscrews into two sections: a T-shaped front section and a long cylinder at the rear. You need just the front section for prime-focus work. This screws into the T-ring, which in turn attaches to the camera, and the whole system is then inserted into the drawtube. If your telescope is an SCT or Maksutov you need a different type of adapter, which is known, rather confusingly, as a T-adapter. The rear threads into a camera T-ring, as before, but the front threads onto the rear cell of the telescope. Meade and Celestron T-adapters work with any of the SCT range produced by these companies, but if your telescope is one of the ETX range you need a special T-adapter made for this line of telescopes.

The prime focus method is fine for whole-disc photography, but with most instruments it does not provide enough magnification to show fine details within the spots and in no instrument does it exploit the full resolving power of the telescope. Also, with some telescopes, such as small refractors and short-focus Newtonians, the image formed at the prime focus is rather small. My 80 mm refractor, for example, has a focal length of 910 mm, giving a prime focus image just over 8 mm in diameter. To magnify the image further there are three somewhat more advanced methods for photographing the Sun.

The Afocal Method

You can obtain a higher magnification by pointing the camera, with its lens in place, into the telescope eyepiece. This is called the *afocal* method and is an excellent way of photographing the Moon and the brighter planets with undriven telescopes that are difficult to attach cameras to, such as Dobsonians. Always use a tripod to mount the camera at the eyepiece, so as to ensure your images are not blurred by camera movement. One major advantage of this method is that the camera is not attached to the telescope and so vibrations produced by the shutter release do not affect the telescope. This is a good method to choose if your camera suffers from heavy shutter vibration or if your filter forces you to use exposures between $\frac{1}{125}$ and $\frac{1}{2}$ second.

The afocal method does, however, have one or two problems. Although it is quite easy to centre and focus the image when you use a low-to-medium power eyepiece, both become difficult at higher magnifications. Another problem is stray light between the camera and the eyepiece. This does not normally cause difficulties with shots of the Moon and other objects taken at night, but during the day stray sunlight will show up even with very short exposures. You can eliminate this with a piece of dark cloth or perhaps a photographic bellows between the camera lens and eyepiece, but remember that the telescope is moving all the time relative to the camera and so will eventually drift out of position! Unless you have no choice, the afocal method is unnecessarily cumbersome, as today good filters, the right camera and a steadily mounted telescope enable other, easier methods to produce good results. Afocal photography is, however, the main method to use with most digital cameras, as I will explain in Chapter 8.

Using a Teleconverter

You can magnify a telescopic solar image quite easily with an ordinary teleconverter lens, as you would use in everyday photography. A teleconverter is especially useful if your telescope's focal length is a little too short to give a good-sized prime focus image. For example, a $2\times$ teleconverter effectively doubles the focal length of my 80 mm refractor (focal length 910 mm), giving it an effective focal length of 1,820 mm, which produces a 16 mm disc size, filling much of the frame. Figure 7.3 shows an example of one of my own photographs taken using this method. Note, however, that as well as

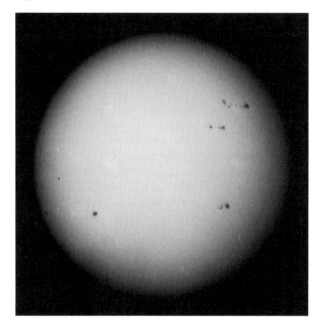

Figure 7.3. Whole disc photograph, taken by the author on 2001 August 8th, using an 80 mm refractor, Baader AstroSolar Safety Film filter at full aperture, and a 2× teleconverter, giving an effective f/ratio of f/22.8 (effective focal length 1,820 mm). Exposure 1/125 second on TP 2415 film.

increasing the effective focal length, teleconverters increase the f/ratio as well – in my case, from f/11 to f/22. Because f/ratios follow a square law, doubling the f/ratio requires an exposure four times as long. For example, if the Sun is correctly exposed at 1/1000 second at prime focus, you will need a $\frac{1}{250}$ second exposure if you use a 2× teleconverter. You can stack two or more teleconverters to give even longer focal lengths, although this increases exposure times even further. An alternative to teleconverters is a Barlow lens, since a teleconverter is a negative lens, like a Barlow. However, Barlows can cause internal reflections. Some otherwise good solar images taken through a low-cost achromatic Barlow of mine were spoiled by such a reflection.

Eyepiece Projection

Another way of increasing the magnification is to use *eyepiece projection*. Here the camera lens is removed and an ordinary telescope eyepiece projects an image of the Sun onto the film – not unlike projecting the image onto a screen for visual observing, although for photography we must, of course, use an aperture filter. If your telescope is a refractor or a Newtonian this is where the second, cylindrical part of the adapter comes in. An eyepiece is inserted into a receptacle inside this section and locked in place with a thumbscrew. This section is screwed to the front part of the adapter and the whole adapter is then threaded into the T-ring (see the diagram in Figure 7.4 and the photograph in Figure 7.5). If your telescope is an SCT you need a *tele-extender*, a simple tube which threads into the T-ring at the rear and into the telescope's visual back at the front. The amount of magnification produced by eyepiece projection depends on the focal length of the eyepiece used and the distance between the eyepiece and the film. Higher magnifications are given by shorter focal length eyepieces, as in visual observing. Simple camera adapters allow you to change the eyepiece-to-film distance by just a few millimetres at best, and then only by sliding the eyepiece in its receptacle. More expensive adapters are available which use two tubes sliding inside each other to vary the projection distance.

The effective f/ratio given by eyepiece projection is slightly more difficult to determine than with other methods, because in addition to the telescope's f/ratio we need to take into account the eyepiece focal length and the projection distance. We can use the following formula to work out the effective f/ratio:

$$\frac{f/_{telescope} \times (\text{projection distance} - \text{fl}_{eyepiece})}{\text{fl}_{eyepiece}}$$

where $f/_{telescope}$ is the telescope's original f/ratio and $\text{fl}_{eyepiece}$ is the focal length of the eyepiece. As an example, suppose you want to get a close-up of a sunspot using an 80 mm f/11.4 refractor. You are using a 7 mm eyepiece and the projection distance, set by the

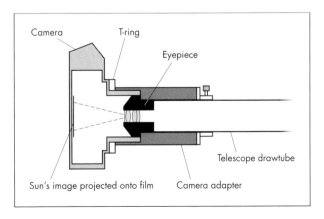

Figure 7.4. Diagram showing the principle of eyepiece projection using a refractor or Newtonian, using a camera adapter and T–ring.

Figure 7.5. Eyepiece projection arrangement on the author's 80 mm refractor, showing Olympus OM-1 camera (with cable release), T-ring and camera adapter.

camera adapter, is 70 mm. The projection distance is measured between the focal plane of the eyepiece and the plane of the film. In practice, it is impossible to measure this exactly, because the focal plane varies from eyepiece to eyepiece and you cannot see the eyepiece inside the adapter. Estimate the distance by measuring from where you judge the mid-point of the eyepiece to be. The position of the film plane should be marked by a small circle with a line through it, engraved in the top or bottom of the camera. Plugging the numbers into the formula, we get:

$$\frac{11.4 \times (70 - 7)}{7} = 102.6$$

We can find the effective focal length provided by this setup by multiplying the new f/ratio by the telescope's aperture. Multiplying 102.6 by 80 gives us 8,208 mm – a long focal length, showing only part of the Sun's disc, excellent for detailed images of individual sunspot groups.

Close-ups of sunspot groups is the best use for eyepiece projection (Figure 7.6). You can also use this method to get whole-disc shots if your telescope has only a short focal length. However, eyepiece projection can cause problems for whole-disc shots. To get the Sun's full disc in the picture you may have to use a low-power eyepiece and a short projection distance, which can cause the field to be "curved" – that is, the centre of the picture is in focus but the periphery is blurred. If the curvature is small you may be able to get away with it, though, as the limb darkening on the Sun causes the edge of the disc to be underexposed, and a small amount of blurring may not be noticeable!

Taking Pictures

Seeing conditions are very important in solar photography. Poor seeing can cause even whole-disc shots, taken at relatively low magnifications, to be blurred, and can render close-ups of sunspots useless. The advice I gave in Chapter 3 on seeing for visual observing applies to photography as well. If you use the projection method for visual work, it is a good idea to do your photography first, as projection allows heat to build up inside the tube, causing turbulence if you do photography afterwards. Thin, high cloud such as cirrus can also dramatically reduce the contrast of solar detail on photographs, even though you can often carry on visual observations in such conditions. If the seeing is poor or significant cirrus is present, you may be better waiting for better conditions before doing photography – unless there is something particularly interesting or unusual to photograph.

Before starting a photographic session, make sure that any eyepieces or other lenses you use, such as teleconverters, are scrupulously clean and free from dust specks. These lenses are often located close to the

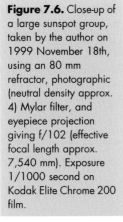

Figure 7.6. Close-up of a large sunspot group, taken by the author on 1999 November 18th, using an 80 mm refractor, photographic (neutral density approx. 4) Mylar filter, and eyepiece projection giving f/102 (effective focal length approx. 7,540 mm). Exposure 1/1000 second on Kodak Elite Chrome 200 film.

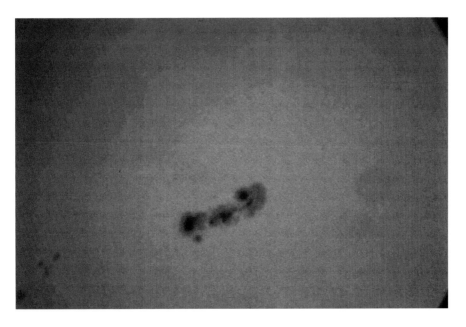

focus of the telescope and dust particles show up on photographs as grey blobs. Dust blobs are not usually noticeable on night-sky pictures, but they stand out very distractingly against the bright background of the Sun and can sometimes be confused with sunspot penumbrae! The majority of dust forms on the exterior surfaces of lenses and you can clean them quite easily using a lens cleaning kit from a camera store. Always use a blower brush (usually provided with the kit) first to remove any large dust particles, as these can scratch the surface of the glass if you try to wipe the lens straightaway. Over the years a small amount of dust seeps into the inside surfaces of the lenses. If a significant amount accumulates you may have to take the lens to an astronomical supplier or a camera repair specialist to have it cleaned professionally. Do not take eyepieces apart unless you know what you are doing, as it is easy to insert lens elements back in the wrong way round.

When you have mounted the camera to the telescope you need to focus the image. As I have explained above, focusing on the solar image can be difficult with the standard focusing screens supplied with most cameras. If you cannot change the screen on your camera, try focusing on the ring surrounding the central spot on the screen, where the image is sometimes brighter and clearer. Avoid focusing on the large outer part of the screen, partly because the image is faint here but mostly because the centre of the image will be blurred if there is any field curvature. Many amateurs find a focusing magnifier a useful aid in getting a precise focus. These devices take different forms. Some of the Nikon cameras have a detachable prism and Nikon's focusing magnifier is a powerful eyepiece which attaches to the top of the camera directly above the focusing screen, when the prism has been removed. The Olympus Varimagni attaches to the viewfinders of models in the OM series and offers a choice of $1.2\times$ and $2.5\times$ magnification. It bends the light through 90 degrees, allowing you to view the image comfortably from above, as with a star diagonal. Other magnifiers are available for other makes of camera, although there are many cameras which do not accept any currently available magnifier. Also, these devices are quite expensive to buy. Whatever you use, if you are uncertain of the correct focus, take several test exposures at different focus positions and mark the positions on the telescope drawtube. When the film is developed you can select the best position for use in the future.

Before starting exposures, lock the mirror or set the self-timer if your camera has these facilities. (Remember that the self-timer is only useful in reducing vibration if it flips up the mirror at the *beginning* of the self-timer operation.) If your telescope has no clock-drive, position the Sun's disc or the sunspot you are photographing a little to the east of the centre of the screen, to allow for movement of the image due to the Earth's rotation in the seconds before the exposure. Use trial and error to determine how far you have to position your target.

Trial and error is necessary for determining exposure as well. Filters, f/ratios and conditions vary so much that it is impossible to suggest precise exposure times. To begin with, take a large range of exposures from $\frac{1}{1000}$ second upwards, making careful notes of the exposure used on each frame of the film (see below). Do not be afraid to try shutter speeds slower than $\frac{1}{125}$ second and see how long an exposure you can use before vibration becomes a problem, as longer exposures are sometimes possible with the right camera and a solidly mounted telescope. Even when you have found the correct exposure, always take an exposure one stop shorter and another a stop longer than the ideal. This is known as "bracketing" exposures and allows for variations in the Sun's brightness due to observing conditions or the altitude of the Sun. In fact, it is a good idea to take *two* shots at each shutter speed, so that you cannot confuse dust particles, scratches or faults on the film with real solar features. From my own personal experience, I have found that with my 80 mm refractor, a Baader AstroSolar visual filter (neutral density 5) and an effective f/ratio of f/33 (achieved by eyepiece projection), $\frac{1}{250}$ second gives me my best whole disc shots on 200 ISO slide film. Using the same arrangement with 100 ISO film or TP 2415 I increase the exposure to 1/125 second. For high-power eyepiece projection close-ups of sunspots, all my successful images have been with exposures of $\frac{1}{250}$ second or shorter with various films and a photographic density (ND4) filter.

A golden role in solar photography (and any other astronomical photography) is always to *take notes*. Only with careful notes made at the time you took the pictures can you determine what is the best combination of exposure, f/ratio, filter and other factors. It is essential to give full details of how you took each image if you send pictures to solar observing organisations or to magazines for possible publication. In addition to the exposure, focal length, f/ratio and filter, for each exposure you should also record the date and time (in UT) at which you took it, together with the observing conditions. It is also important to note what method you used (e.g. eyepiece projection or prime focus), any means of reducing vibration and any problems encountered.

When the films have been processed, you should store them carefully, in order to preserve your precious images and also so that you can compare them easily with the written record you have kept at the telescope. When I receive my slides back from the laboratory, I cut and mount each frame myself, labelling the mount with the frame number and the month and year on which the film was developed. Store slides in files with plastic "pages" containing clear pockets holding one slide each, to protect them from dust and long exposure to light and air. You can store negatives in similar file pages with slots for strips of five or six 35 mm frames.

Photography in Hydrogen-Alpha

Photographing the chromosphere and its features through H-alpha filters is more difficult than white-light photography, because much less light is available. It is impossible to take successful photographs in H-alpha with very long effective focal lengths as you can in white light, unless you have a large telescope and a correspondingly large (and very costly) filter, because exposure times are very long and the resulting images will be blurred. The main method for H-alpha photography is prime focus.

Your choice of film is very important in H-alpha photography. The H-alpha line is located well into the red part of the Sun's spectrum, which means that your film must have good sensitivity to red light. Not all films have this property. The only suitable black and white film is Kodak 2415, which has a very strong response to red light. Most other black and white films have very poor red sensitivity. For colour photography of prominences I have found Kodak Elite Chrome 100 and 200 to work very well, but the equivalent films made by Fuji (i.e. Fujichrome Provia and Sensia) currently have a poor red response and so are less suitable. However, manufacturers sometimes change the characteristics of their films and so it may be a good idea to try out these and other films occasionally. A good test of a film's suitability for H-alpha work is its performance in deep-sky photography at night. Try photographing one of the brighter nebulae, such as the Orion Nebula, M42, or M8 in Sagittarius. You can do this very simply by taking exposures of up to a minute's duration with your camera on a fixed tripod. Emission nebulae such as these emit strongly in H-alpha, which is why they appear red on photographs. (You cannot see this red colour visually because your eyes – like some films – have poor red sensitivity). If the nebula appears bright and has a rich red colour with a short fixed-tripod exposure then the film has good red response and is suitable for H-alpha photography. The film is not suitable if, after the same length of exposure, the nebula appears faint and grey.

If you have a coronagraph you can obtain spectacular images of prominences, because the Sun's glare is blocked by the occulting disc (Figure 7.7). The Baader instrument, attached to my 80 mm refractor, gives an effective f/ratio of f/17 – quite fast compared to many H-alpha systems – allowing very short exposure times. I have found that $\frac{1}{250}$ second on Elite Chrome 200 and $\frac{1}{125}$ second on Elite Chrome 100 or TP2415 give the best results, although exposures sometimes need to be adjusted for especially bright or faint prominences. The effective focal length of the system is 1,365 mm, which gives a good disc size on the film and allows even large prominences to be accommodated on the frame. The coronagraph has a diaphragm for controlling the amount of scattered light in the system and I have found that stopping this down about half-way gives the highest contrast. The main problem with coronagraph photography is keeping the Sun hidden behind the disc. This demands accurate polar alignment, which is not easy in the daytime, when you cannot see Polaris or any other stars. This presents no difficulty if your telescope is permanently set up, but if the instrument is portable you need to find another way. One possibility is to align the mount in the normal way at night and make marks in the ground showing the correct positions for the tripod legs – assuming no one minds marks being made on the lawn! I prefer to use trial and error, first aligning the telescope roughly north–south using trees and other local landmarks and then adjusting the mount until the Sun no longer slips out from behind the disc.

A very useful accessory for the Baader coronagraph is called the VIP Excenter. This enables you to move the camera (or eyepiece) at right angles relative to the instrument's optical axis, allowing you to use higher magnifications. (Increasing the magnification makes the Sun – and the occulting disc – larger than the field of view and so the prominences would be out of view if you used higher magnifications with a fixed eyepiece.) The device enables you to take close-ups of prominences using a teleconverter (Figures 7.8–7.9). Remember, though, that doubling the focal length increases the exposure time by a factor of 4, so the $\frac{1}{250}$ second exposure on Elite Chrome 200 mentioned above would need to be increased to $\frac{1}{60}$ second. I prefer not to lock the camera mirror when taking photographs with the coronagraph, as manipulating it with the fingers can move the telescope slightly and so spoil the precise alignment of the occulting disc with the Sun.

Most amateurs use TP2415 film for H-alpha disc photographs. The chromospheric network and most H-alpha activity on the disc has a delicate appearance and its contrast is low, so only TP2415 has enough contrast to show these features well. Exposures using DayStar filters are about the same regardless of the size of your telescope, because most of them operate at f/30. Most amateurs have found that an exposure of $\frac{1}{30}$ or $\frac{1}{15}$ second is required for disc features and $\frac{1}{2}$ second or more may

Figure 7.7.
Prominences photographed by the author on 2001 February 24th, using a Baader coronagraph attached to an 80 mm refractor. Exposure 1/125 second on Kodak Elite Chrome 200 film.

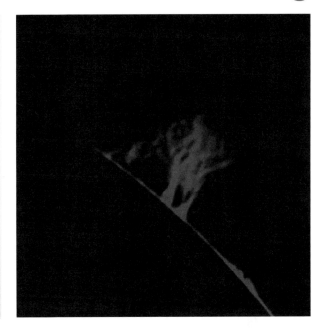

Figure 7.8. A prominence photographed by Derek Hatch on 1998 July 13th at 0657 UT, using Astro Physics 150 mm (6 inch) f/9 refractor, a Baader coronagraph, and stacked teleconverters magnifying the prime focus image by 6×. Exposure 1/15 second on Fuji Sensia 400 film.

be needed for prominences. It may be possible to use the hat-trick method to photograph prominences, but exposures for the disc fall between $\frac{1}{125}$ second and $\frac{1}{2}$ second, and so are prone to vibration. Use a very steady mount and a camera with a very smooth shutter release

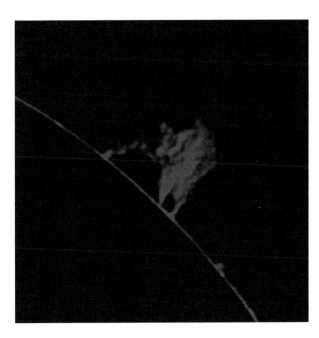

Figure 7.9. The same prominence as shown in Figure 7.8, again photographed by Derek Hatch, this time at 14:21 UT on the same day. Note how the structure of the prominence has changed in the intervening 7½ hours. Equipment, film and exposure are the same as for Figure 7.8

and mirror movement. If you have a Coronado filter, exposure times can vary considerably, because these filters are mounted at the front of the telescope and can be used on many different instruments. If you photograph prominences through any sub-angstrom filter, whether it is made by DayStar or Coronado, the Sun's disc will be overexposed and may appear distractingly bright. You can get round this by masking out the disc during the processing stage, using either a suitably shaped piece of paper above the print on the enlarger easel or by scanning the image and using digital processing techniques. The resulting image then simulates the effect of a coronagraph.

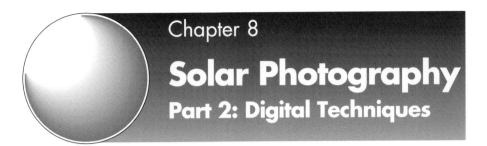

Chapter 8

Solar Photography
Part 2: Digital Techniques

Digital photography is nowadays an efficient way for the amateur to record sunspots and other solar features. Digital cameras are readily available off the shelf at camera stores everywhere and nearly all models can be used at least to some extent for solar photography. There are many excellent cameras well within the amateur astronomer's budget and prices are dropping all the time. At the time of writing (late 2001), some entry-level models were selling for £100 or less and many cameras capable of taking really good solar images were available for under £500.

From the viewpoint of the amateur solar observer, digital photography has many advantages (Figure 8.1). The most important is that it produces instant results. Once you have taken a picture, you can see it immediately on the camera's LCD monitor. If you don't like the picture, you can erase it and try again with another shot. With a conventional camera you have to shoot a whole roll of film and have it processed before you can see the results. Because digital photography allows you to view the results immediately, you can learn from your mistakes much more quickly. A second advantage of digital photography is that you do not need a darkroom to process the images or have a laboratory do the work for you. All you need to process, handle and print out the images is a standard personal computer – something which many people, especially amateur astronomers, either possess or have regular access to nowadays. Because the images are produced

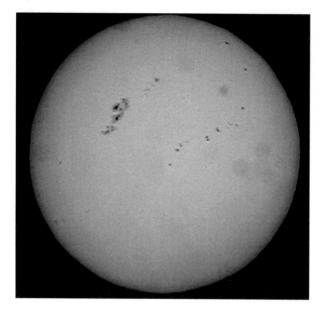

Figure 8.1 An example of what is possible with today's digital cameras. A whole-disc image taken by the author on 2001 March 29th using an Olympus C-860L 1.3 megapixel digital camera *hand-held* to the telescope eyepiece! The telescope was an 80 mm refractor with Baader AstroSolar Safety Film filter and a low–power eyepiece in place.

on a computer, they are ready for sending as e-mail attachments or posting on the World Wide Web. A daily solar image taken with your digital camera would be a great addition to an astronomy club Web site, as would be a "movie" of many solar images taken at daily intervals showing the Sun's rotation. This capability of producing instant results and sending them by e-mail makes digital photography very useful for reporting interesting or unusual solar events to solar observing organisations. Digital images can be quickly distributed to other members of these groups. If you are ever lucky enough to see a white-light flare, digital images would be especially valuable. If your digital camera and telescope are aimed at the Sun when such a flare is in progress, and if you e-mail it quickly enough to your observing group, your image might be one of the first pictures of the event available.

As well as the everyday digital camera, there is also the CCD camera specifically designed for astronomical imaging through telescopes. These cameras have revolutionised night-sky imaging, because their high sensitivity greatly reduces exposure times, enabling amateurs to capture very faint detail in deep-sky objects and to take very high-resolution shots of planets. But for solar photography in white light this

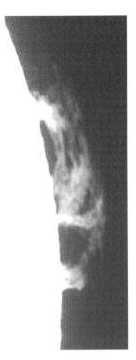

Figure 8.2 CCD image of a hedgerow prominence, taken by Eric Strach on 2001 May 24th at 11:33 UT, using a 200 mm (8 inch) SCT stopped down to 63 mm (2.5 inch), DayStar 0.6 angstrom passband filter and Starlight Xpress SXL8 CCD camera. Equipment the same in Figures 8.2, 8.3 and 8.4.

sensitivity has little advantage, as the Sun already has plenty of light. Secondly, due to the small size of the light-sensitive chip in most CCD cameras, only part of the Sun is visible on one frame, unless your telescope's focal length is very short. They are also more expensive than most consumer digital cameras, and are less convenient to use, as most models do not have a built-in LCD monitor but need to be hooked up to a computer while the images are being taken. If you do not have an observatory such an arrangement is a chore to use, as you have a lot of equipment to set up and take down each time you want to image the Sun. (One notable exception is the STV produced by Santa Barbara Instrument Group, whose control unit has an optional 130 mm (5 inches) LCD monitor and the capacity to store up to 14 CCD images without the need for a computer.) If you have a CCD camera, however, you can use it to take excellent images of the Sun (Figures 8.2–8.4). One area in which CCDs score over traditional photography is H-alpha imaging. The high sensitivity of the CCD chip allows you to capture detailed images of H-alpha disc features with short exposures.

You do not even need a digital camera to benefit from digital techniques. You can also scan conventional slides, negatives or prints into computer-readable format and then enhance them with image processing software just as easily as images taken with a digital system. There are several methods for scanning your images. The best is to buy your own scanner and do it at home. For prints an ordinary A4 flatbed scanner, available for well under £100 from any computer store, is fine. To scan slides or negatives you need a film scanner, which is more expensive, although these devices are coming down in price all the time. A cheap and popular option is to have your 35 mm images scanned onto a CD-ROM, such as Kodak's Photo CD. Many photographic laboratories and even high-street camera shops have facilities to do this on the spot, without the need to send original slides or negatives through the post. Many professional laboratories can also scan images onto CD-ROM, often at very high resolution, but they can be expensive. Depending on the resolution the images are scanned at, you can store up to 100 images or more on a CD-ROM.

Equipment for Digital Photography

Cameras

If you already have a digital camera, you can probably use it to some extent for solar photography. The one essential feature is a built-in LCD monitor both for composing the picture and viewing the resulting image. Most digital cameras are not of the SLR type but work like 35 mm compact or "point and shoot" cameras. With these cameras you do not look through the lens but use an entirely separate viewfinder to frame your shot. This viewfinder is located in a different position to the lens which takes the picture, which is no problem in everyday photography, but if you are shooting through a telescope this prevents you from seeing what the lens is seeing!

If you are choosing a camera specifically for solar photography, there is a huge range of models available in the camera stores, with prices ranging from tens through to thousands of pounds. However, certain features make some cameras much more suitable for solar photography than others. The most important is the camera's resolution. Like a dedicated CCD camera, a digital camera uses a light-sensitive CCD (charge-coupled device) chip in place of the film. This chip – and the resulting picture made from it – is composed of myriads of tiny elements called pixels. The larger the number of pixels on the chip, the higher its resolution, in the same way as the resolution of photographic film is higher if it has fine grain. The resolution of a digital camera is measured in megapixels – i.e. the number of pixels divided by 1 million. So if you see a camera advertised as having 3.3 megapixels, its CCD chip contains 3.3 million pixels – more than enough for quality solar imaging. In fact, any camera with a resolution of more than 1 megapixel is adequate for solar photography. This requirement really rules out the very cheap low-end cameras selling for well under £100, as their resolution is generally less than 1 megapixel. Similarly, Webcams may be satisfactory for first experiments, but their resolution is again too low for serious work. Generally, a good first camera will be in the 1 megapixel range, costing around £200 ($300) or a little less, but if your budget stretches to £500 ($800), you may wish to consider a 3-megapixel model.

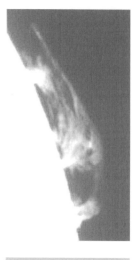

Figure 8.3 CCD image of the same prominence as shown in Figure 8.2, taken by Eric Strach at 12:55 UT on 2001 May 24th, showing changes in the prominence's structure since the previous image.

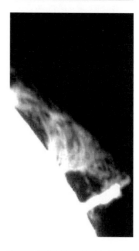

Figure 8.4 The same prominence as that shown in Figures 8.2 and 8.3, this time taken on 2001 May 25th at 14:08 UT, again by Eric Strach.

Everyday digital cameras store images on memory cards – small, removable units slotted into the camera. It may be worth purchasing extra cards to supplement the standard card that comes with the camera, as each image takes up a large amount of memory. For example, the 8-megabyte (Mb) card that comes standard with my Olympus C-860L camera (resolution 1.3 megapixels) will store only two images at full resolution. Consider getting at least 64 Mb worth of memory. As with RAM on personal computers, you can never have too much! On the other hand, as I will explain below, you do not always need to use the camera's full resolution, so the standard card may be fine to begin with.

A third useful feature is a reasonably large lens. Most digital cameras do not have removable lenses, so you need to choose the lens when you choose the camera. Your choice of lens is particularly important in solar photography, because you will be using the afocal method – i.e. pointing the camera lens into the telescope eyepiece. A small camera lens with a short focal length shows the telescopic field as very small and surrounded by blurred telescope parts – a phenomenon known as vignetting. Another problem caused by small lenses is that they show only part of the telescopic field, so that even with a low-power eyepiece part of the Sun's disc is obscured. Such a shot is obviously rather unattractive. A longer lens excludes most of the surroundings and shows more of the telescope's field of view, allowing more aesthetically pleasing pictures showing just the Sun surrounded by black sky. Even better is a camera with a zoom lens or telephoto facility.

Because the camera is not attached to the telescope in the afocal method, and also because digital cameras do not use moving parts when the shutter is released, vibration does not pose the same problems in digital photography as it does with conventional cameras. However, vibration can still cause trouble, especially if (as described above) the camera has to be precisely aligned to get the Sun in the picture. The problem is made worse by the fact that, because they are designed for everyday amateur photography, many digital cameras have no facility for a cable release and so you have to squeeze the shutter release by hand. Fortunately, digital cameras often have a self-timer facility, which gives an interval of several seconds for any vibrations caused by the hand to die down before the shutter is opened. Try to make sure a camera has this feature before buying it. Even better is a remote control handset which you can use to operate the camera from several feet away, but only the more expensive models have this feature.

Very few digital cameras, unfortunately, have removable lenses and can be attached to telescopes using T-rings and adapters in the same way as a 35 mm SLR. Digital SLRs with removable lenses do exist. They are made to fit lenses produced by major camera brand names and so suitable T-rings for attaching them to telescopes are easy to obtain. A few amateurs have obtained spectacular results with these cameras, but they are primarily intended for professional photography and their prices start at around £2,000 or so. Do not assume that a camera advertised as a digital SLR has a removable lens. One such camera I enquired about turned out to have a permanently attached zoom lens, but it was technically an SLR because, unlike most digital cameras, it had through-the-lens viewing.

Power

Unlike 35 mm solar photography, in which some of the best cameras are fully mechanical, digital work requires battery power. Digital cameras are particularly problematic in this regard, because their built-in LCD screens consume a lot of power. This is not a major problem in everyday photography, as you can keep the screen switched off when composing and taking the shot and use the viewfinder instead. You need only turn the screen on briefly to inspect the picture. But as I explained above, when shooting through a telescope you need to use the LCD screen all the time and so even a short imaging session can completely drain the batteries. To avoid having to constantly replace batteries, buy a battery re-charger and re-chargeable batteries when you purchase the camera. If possible, use nickel metal hydride batteries, as they last somewhat longer after each charge than do nickel cadmium cells. It is a good idea to obtain two sets of batteries and keep them fully charged, so that if one set runs out in the middle of a session you can still keep going. Many cameras have a facility for an adapter to allow the camera to run on mains electricity, but I do not recommend this for using a camera outdoors, as it often involves trailing a live cable over damp grass and perhaps metal telescope or tripod parts.

Mounting the Camera

Because most digital cameras do not have removable lenses, how to position the camera to the eyepiece requires some thought. You can get quite good results by simply holding the camera up to the eyepiece, but you may well have problems with camera shake or partly obscured images caused by the camera being not quite squared on to the eyepiece. It is much better to mount the camera on a separate tripod, which both avoids vibration and allows you to precisely centre the camera over the eyepiece using the tripod adjustments. A number of suppliers advertising in astronomy magazines have produced brackets that allow various models of cameras to be clamped to the telescope. Some cameras, particularly the more up-market models, have lenses with threaded barrels and some suppliers sell adapter rings for attaching these cameras to an eyepiece or camera adapter. Before buying a camera you may wish to enquire as to whether a suitable adapter is available for it.

The Computer

To use a digital camera, you need a computer and suitable software for processing the images. The computer requirements are quite simple. Any computer made in the late 1990s or later is adequate for processing and handling digital images, although the older the machine the longer it will take to do the job. If you use a PC, it should have a Pentium-class or higher processor. If you have a Macintosh you may find that your choice of camera is restricted, as many popular digital cameras are only compatible with Windows. The most important computer requirement is RAM memory, as image files can be very large. To handle images efficiently you ideally need 64 Mb or more if possible. The "system requirements" specified on the software that comes with the camera is a good rough guide as to how much computer power you need. You can store your images on the computer's hard disc to begin with, but the large size of image files could cause it to fill up quickly. In the long run, it is a good idea to invest in a Zip drive or a CD recorder, both of which can store large volumes of data.

The Telescope

Like 35 mm cameras, digital cameras can be used with any kind of telescope, provided it is equipped with a safe aperture filter. If your telescope is a Newtonian, digital photography may be the best method for obtaining pictures of the Sun's whole disc, as the afocal method does not require excessive focus travel as does prime focus photography. Use the same aperture filters as for visual observing. Digital cameras have quite high sensitivity, so you should not need a neutral density 4 photographic filter unless you are shooting at very high magnification. As in 35 mm photography, a motor-driven mount is not essential, although it is certainly useful, as when you are lining up and focusing the camera you do not need to worry about the Sun's apparent motion through the field.

Taking Digital Images

To take solar images with a digital camera, begin by mounting the camera on the tripod or bracket you have chosen. Switch on the LCD and ensure that the camera is aimed squarely into the eyepiece. If you are using a tripod, and if your telescope is a refractor or a catadioptric, turn the star diagonal round so that the eyepiece is in a horizontal position. This makes positioning the camera much easier than when the diagonal is in its usual upright position. Also, place the camera a centimetre or so away from the eyepiece to allow the drawtube to be racked out for focusing. The camera's distance from the eyepiece does not affect the focus in afocal photography, so you can move the camera closer in once you have finished focusing.

Use a low-power eyepiece to begin with. Higher powers have less "eye relief" – that is, you need to place your eye or camera lens closer to it to see the full field of view, which makes it difficult to align the camera. Choose a magnification which shows the whole solar disc comfortably within the field of view. Avoid using very low powers, as they show the Sun's disc surrounded by a lot of black sky. Most digital cameras have an automatic exposure facility, which determines the exposure by the average brightness of the picture. Too much black sky can give a low average brightness and so cause the camera to overexpose the Sun's disc.

Because they were designed for point-and-shoot photography, many everyday digital cameras have no facility for setting the shutter speed (or the aperture) manually. Digital cameras do have exposure compensation and spot metering facilities, but to begin with it is simpler to use automatic exposure. One very good way of increasing the magnification enough for the Sun to fill most of the frame is to use a low-power eyepiece and a Barlow lens, which boosts the power but does not sacrifice eye relief. Using a Barlow lens with a medium-power eyepiece is also the best way to achieve high magnifications when you are ready to go on to more advanced work, as you still have some eye relief for aligning the camera. Alternatively, use the zoom facility on your camera's lens, if it has this. Do not use "digital zoom", however, as this merely increases the size of the pixels, leading to a grainy picture. Genuine zoom is sometimes known as "optical zoom". As with film photography, it is important to use clean eyepieces and other lenses, as dust particles can show up as distracting grey spots on the image.

The LCD screen can be hard to see in full sunlight, especially if you are using a Newtonian or a star diagonal and the Sun's light is striking it sideways-on. To make it easier to see, make a simple shade from black card and stick it to the camera with tape. It is also important to block out, or at least minimise, stray light between the camera lens and eyepiece. Wrap a dark cloth loosely around the camera and drawtube, ensuring it does not prevent you from using the camera's controls, or make a short tube from black paper so that it hangs loosely over the eyepiece and lens barrel and bridges the gap between them. Make sure it is not too tight, however, to avoid transmitting camera vibrations to the telescope and to allow for minor adjustments of the camera's position.

You now need to ensure that the image on the LCD screen is focused. Most digital cameras have automatic focusing – commonly referred to as autofocus – and, as with exposure, many popular models have no means of overriding this and setting the focus manually. Fortunately, many digital cameras focus quite accurately on the solar image in the eyepiece, as long as you have focused it carefully with your eye beforehand. (If you wear glasses, it is important to focus the telescope with them on, as without them the focus will be inaccurate.) Many observers find it hard to determine whether the image on the LCD is precisely in focus, not only because stray sunlight makes the screen look dim but also due to the small size of the image and the screen's rather low resolution. If your camera has a video output socket, you could connect it to a portable TV and so view the image on a larger and brighter screen. A number of amateurs have used this method successfully, but if your equipment is portable it is inconvenient to have to set up a TV each imaging session in addition to the telescope and camera. If you have any doubt as to the focus of the picture, it may be better to use trial and error in your first imaging sessions, by marking the telescope drawtube with several focus positions and taking a series of pictures with the drawtube in a different position each time. If you make a note of the focus position for each picture, you can establish which position gives the best focus when you view the images on the computer later.

Although the camera's shutter speed is set automatically, you can control the brightness of the image to some extent using two features found on most popular digital cameras: the ISO rating and exposure compensation. The first feature allows you to set the camera's sensitivity to light based on the ISO system of rating film sensitivity. On my own camera, I have found that the slowest setting, 125 ISO, is quite adequate for solar photography, as the Sun's disc is a bright subject, even when imaged through an aperture filter. Also, higher speeds give a more grainy picture, just as with film. Only if you are shooting at high magnification should you select a faster speed to keep exposure times down.

A final important setting is the resolution of the image. Although the highest resolution gives the best quality picture, it is better to select a medium resolution setting to begin with or if you are experimenting with new techniques. High-resolution images take up a lot of memory. For example, the 8 Mb card on my 1.3 megapixel camera will store only two images at the camera's highest resolution, whereas it will hold 36 medium-resolution images – the equivalent of a roll of 35 mm film. Taking a large number of medium-quality images allows you to experiment widely without changing cards and to download the results to the computer more quickly. Moreover, you will be pleasantly surprised at how much sunspot detail you can record even at medium resolution.

When you have aligned the camera, focused the image and adjusted the various settings, you are ready to expose. Squeeze the shutter release button as gently as you can.

Some digital cameras require you to press the shutter release quite hard and the camera can move by a surprising amount, even when it is mounted on a separate tripod. Again, always use the self-timer facility or remote control if your camera has either of these features.

Be prepared for a high failure rate to begin with, as there are so many things you have to get right in digital photography with the afocal method – controlling stray light, aligning the camera, magnification, focusing, exposure and camera shake. In my experience, the main difficulty in digital photography is maintaining camera alignment during the exposure, especially if the camera is mounted separately from the telescope. Because the available field of view is often restricted, even a small movement of the camera, such as that caused by pressing the shutter release, is enough to jog the camera so that the Sun is half-hidden. Even the self-timer does not always compensate for this. But digital photography, with its instant results and the ability to throw away an image there and then if you don't like it, allows you to experiment to your heart's content. You can take 10 or 20 "dud" images and then have the satisfaction of getting a good one. Don't be too hasty about throwing away images, though, unless they are really bad, as if it looks reasonably well-centred and focused you can usually do something with it at the image processing stage. As I shall explain below, it is surprising what you can do even with a relatively poor image.

As with film photography, always take notes on your images, recording details such as the eyepiece and Barlow lens used, the ISO setting and any exposure compensation, as well as the date and time. The latter two details are recorded automatically on most cameras, but they sometimes return to the defaults when you change the batteries and it is easy to forget to reset them.

Handling and Processing Digital Images

You can transfer (or "download") the images to the computer using the equipment provided with the camera. This often takes the form of a cable connecting the computer to the camera itself, and the images are downloaded with the memory card still in the camera. An alternative is to slot memory cards into a special card reader and download them to the computer from there. Most cameras save digital images in standard formats such as JPEG or TIFF, so you can view them with any image-processing or image-editing program.

When you first see them on the computer screen, the "raw" images fresh from the camera may appear disappointing, with sunspots looking washed out and the whole picture somewhat fuzzy. But don't be discouraged, as such images contain a lot of data which you can extract using the computer. This is where image processing comes in (Figure 8.5). As I have explained above, image processing is not just for images taken with digital cameras. The techniques described in the rest of this chapter apply equally to scanned images originally taken on film.

Software for processing images varies greatly, both in price and the number of features it contains. The brand-name image processing packages used by professional photographers and publishers are very expensive. For example, the current version of Adobe PhotoShop retails for over £500. But many programs containing most of the features you need for processing solar images are available very cheaply. Many digital cameras come with excellent image processing software – sometimes brand name programs – which contain most of the features you need. In fact, the image processing software is an important factor to consider when choosing a camera. Good packages are also frequently "bundled" with new computers, printers and scanners. One of my favourite image processing programs is Adobe PhotoDeluxe, which came free with an inkjet printer. Even Microsoft Photo Editor, which comes free with some Microsoft programs, has enough basic image enhancement features for you to get started in digital processing. If you are contemplating buying a software package, either on its own or with other equipment, check that it has the following features:

- Ability to trim or "crop" the picture to exclude unwanted material;
- Ability to flip the image horizontally or vertically;
- Individual controls to adjust the brightness and contrast of the image;
- "Instant fix" feature to improve brightness and contrast automatically;
- "Sharpen" feature;

Figure 8.5. How a seemingly poor digital image can be greatly improved using image processing. **a** Raw image, obtained by the author on 2001 August 24th using the same camera and telescope setup as in Figure 8.1, except that this time the camera was mounted on a tripod. **b** The image after having been "cropped" to exclude the worst parts of the picture and then undergoing moderate contrast enhancement and unsharp masking.

a

b

- Colour hue and saturation adjustments;
- Unsharp mask.

If a program has most or all of these features, then it is adequate for getting started in processing solar images.

Before beginning to process images, it is essential to remember one golden rule. When you have chosen an image you want to enhance, always work with a copy of the file, **never** the original image. This precaution ensures that you can always return to the original image if things go wrong during processing. It is not safe to rely on the "Save Changes?" dialogue box when you close an image, as it is only too easy to click on "Yes" by mistake and so lose your original image forever.

To demonstrate how the various features of an image processing program can improve a solar image, I shall use as an example one of my own solar images, taken on 2001 October 18th with an Olympus C860L 1.3 megapixel camera and showing the whole solar disc.

Cropping and Trimming

Figure 8.6a shows the original image, looking washed out and with the sunspots barely visible. The Sun's disc is off-centre and there is rather a lot of black sky around the disc, which may partly explain why the sunspots are washed out, because the average brightness of the picture is too low and so the camera has used too long an exposure. The first thing we can do to make the image more aesthetically pleasing is to trim the edges of the photograph to show the solar disc in the centre of the picture and filling most of the frame. Selecting the "crop" or "trim" feature causes the mouse pointer to turn into a symbol representing a cutter or a pair of scissors. Position the pointer in the top left corner of the picture, then click the mouse and "drag" it down and to the right. The selected area of the image appears as a rectangle outlined by dotted lines, whose shape varies as you move the pointer; dragging it to the right gives a horizontal rectangle, while dragging it down increases the vertical height of the image. When you have selected the area you want to include in your final image, release the mouse button. At this point the program will ask you to confirm your selection. The selected area then fills the available area on the computer screen. Figure 8.6b shows the result of cropping my example image. Of course, cropping an image enlarges it slightly, increasing the size of the pixels – the digital equivalent of film grain – but, even though I took the original image with just the medium resolution setting on the camera, the pixels are not large enough to be noticeable. Cropping also allows you to exclude parts of an image that are blurred or otherwise sub-standard and detract from the quality of the picture.

Image Orientation

An image taken with a digital camera using the afocal method has the same orientation as the visual view through the eyepiece. However, the orientation of the image on your computer screen may be different from what you are used to seeing, especially if you normally use the projection method, which reverses the east–west orientation. If you shoot through a Newtonian or a refractor without a star diagonal, north and south will be reversed as well. Turning the eyepiece round to a horizontal position in a refractor or catadioptric (see "Taking Digital Images" above) also changes the orientation. Image processing allows you to quickly and easily turn the image round so that it matches drawings and other photographs. Amateur drawings use a standard orientation showing north at the top and east to the right. If you send images to a solar observing organisation they should have this orientation.

To change the north–south and/or east–west orientation of your image, use the "Flip Horizontal" and "Flip Vertical" commands. Figure 8.6c shows the example image flipped vertically to show north at the top and horizontally to show east to the right.

Changing the Brightness and Contrast

How well the sunspots and other features show up on any image depends on the contrast. Getting the contrast exactly right on a film image can involve hours of patient experimenting in the darkroom with different chemical formulae, tricks with the enlarger and printing techniques. Digital image processing dispenses with all that and allows you to find the correct contrast very quickly. Many image processing packages have a feature called "Instant Fix", which adjusts the brightness and contrast of an image automatically. While this feature certainly enhances the visual impact of an ordinary photograph, it is often not adequate for solar images, which need much greater adjustment. It may be sufficient if your image already shows solar details with good contrast but is a little too dark.

Most solar images require use of the "brightness and contrast" feature. In most image processing packages this takes the form of a box with individual scales for

a

b

Figure 8.6. Stages in improvement of a "raw" image using image processing. **a** Raw image, taken by the author on 2001 October 18th using the same equipment setup as in Figure 8.5. **b** The image trimmed to centre the Sun's image and exclude some of the black background. **c** The image flipped to show north at the top and east to the right.

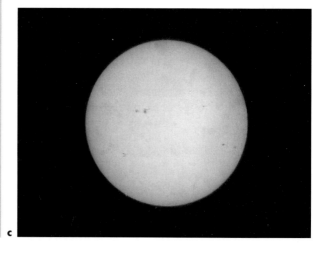

c

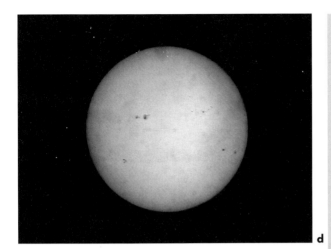

d

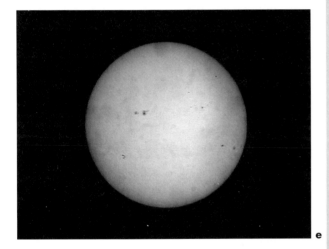

e

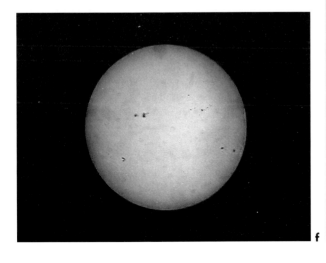

f

d Brightness reduced by 85 points and contrast increased by 50 points.
e Image sharpened using the "Sharpen" feature.
f Contrast and sharpness enhanced further using unsharp mask.

brightness and contrast, running from –10 to +10 or –100 to +100. You can set the brightness and contrast either by entering numbers between these values in the boxes provided or by sliding a pointer along the scale using the mouse. Many packages have a "preview" box in which you can see the effect of any changes "live" without having to close down the menu item, and some programs allow you to preview the entire image in the background without committing to it. It is interesting to play around with the brightness and contrast scales to see the effect of changing the values. Note that increasing the contrast value alone does not improve the contrast of the sunspots; in fact, it makes the image look brighter and *reduces* the contrast of the spots. If your raw image is overexposed, like my example image, to make the spots show up better you must first decrease the brightness value substantially and then increase the contrast by a somewhat smaller amount. In Figure 8.6d I have decreased the brightness value by 85 points out of a possible 100 and increased the contrast by 50 points. The sunspots are now much darker and the image now shows several smaller spots that were not visible on earlier versions.

Changing the Colour

As I have noted in previous chapters, the colour of a solar image is not important, as the Sun is really a monochrome object. However, it is sometimes useful to be able to change the colour of an image to make it more aesthetically pleasing, as some filters give the Sun a lurid colour. Some Mylar filters, for example, give the Sun a strong blue tint while many glass filters make it look unnaturally orange. With most image processing packages it is easy to change the colour of an image using the "Colour Balance" feature. This contains three sliding scales, similar to the ones for brightness and contrast, running from cyan to red, magenta to green and yellow to blue. The latter scale is especially useful for reducing or eliminating the blue tint of a Mylar image. Moving the pointer slightly towards the yellow end of the scale makes the blue colour paler and less harsh. Moving it further gives the image an almost neutral, monochrome shade and moving it further still gives a pleasing pastel shade of yellow. If your filter gives a strongly orange image, you can get a more pleasing colour by moving the pointer towards the blue. A Baader AstroSolar filter sometimes gives the image a slight purple shade, which can be corrected by moving the pointer slightly towards the green on the magenta–green scale. The third scale, running from cyan to red, is useful for correcting any reddening of the image caused by the Sun's low altitude. Simply move the pointer towards the cyan end to restore the image's normal hue. If you are happy with the actual colour of your original image but would like it to be less intense, most packages allow you to reduce the colour saturation, again using a pointer on a sliding scale. Reducing the saturation by the maximum amount effectively turns your image into a black-and-white picture. Some programs also have a "colour to black-and-white" feature that allows you to turn the image to monochrome at the click of a mouse.

Removing Dust and Scratches

I noted above that dust particles on the telescope eyepiece can show up as unsightly grey spots on the image. In theory, digital processing allows us to remove these. Using the "Remove dust/scratches" feature on an image processing program, you can erase dust spots above a specified size and intensity. But this feature can cause particular problems for solar photographers. It removes the dust particles all right, but it can also remove all the sunspots! It also has the effect of blurring the image. With care this feature can give some improvement, but on the whole I do not recommend it, and prefer to use clean eyepieces to begin with. If you have one or two particularly bad dust spots, it is possible to "paint" them out by selecting the "brush" feature and painting over the spots with the same colour as an unspoiled region of the image. You must get the brightness and colour exactly right, however, as even a slight error can give a ghastly blotch on your image, much worse than the original dust spot. (This demonstrates the importance of only working with a copy of the original image.) In any case, always note carefully where you have altered parts of the image and with what technique, as an image that has been altered in parts is no longer scientifically accurate.

Sharpening the Image

Our image (Figure 8.6d) now looks far better than the original, but it is still not quite as sharp as it could be. You can correct minor blurring such as this at the click of a mouse by using the "Sharpen" feature. Using "Sharpen" just once is enough to noticeably improve the focus of the picture, especially at the limb of the Sun, as Figure 8.6e shows. Sharpening does not increase the resolution of the image, but just makes the detail stand out better – you can't resolve detail that is not already there! Applying "Sharpen" more than once improves the focus even more, but the image becomes much more "grainy", detracting from the overall quality of the picture.

It is possible, however, to extract even more information from a digital image using an image processing feature known as "unsharp mask". This allows you to enhance the contrast and sharpness of the image further while controlling the amount of "grain" (technically known as "noise" in digital imaging) in the picture. Selecting "Unsharp Mask" in your image processing package brings up a box with three sliding scales named "Amount", "Radius" and "Threshold". "Amount" runs from 1 to several hundred percent, "Radius" from 0.1 to 250 pixels and "Threshold" from 0 to 255 levels. Setting the amount to a high level makes the picture look sharper, but at the expense of excessive image noise. "Radius" controls the contrast: selecting a high radius makes the image look bright and stark, especially towards the limb, while reducing it lowers the contrast. "Threshold" makes the image smoother (i.e. reduces the amount of "noise") when set to a high level but lowers the contrast, while a low threshold gives a noisy picture and makes the contrast too high, enhancing every dust particle and showing the sunspots surrounded by bright rings, which could be mistaken for faculae!

Clearly, we need to find a compromise between contrast, sharpness and noise using the three controls. I have produced my best results by setting "Amount" to a medium value (around 200 per cent), "Radius" to between 3 and 20 pixels and "Threshold" to a fairly low value, around 25 levels. Figure 8.6f shows our example image, which has already been through "sharpen" and has had its brightness and contrast adjusted as described above, treated with the unsharp mask feature with the amount set to 200 per cent, the radius left at its default value of 5.6 pixels and using a threshold of 25 levels. The result has even higher contrast and shows more detail, but without increasing the noise too much or causing facula-like rings around the spots. Note that several more small spots have come "out of the woodwork".

Unsharp mask could actually be used to *reduce* the contrast if it is too high in the original image. This could improve images taken on Kodak 2415, whose inherent high contrast sometimes exaggerates the limb darkening and causes features near the limb to be underexposed.

Appendix A
Equipment Suppliers

This list is not intended to be fully comprehensive but rather is a selection of suppliers of solar observing equipment that I and other amateurs have found to be useful. It does not constitute a recommendation of these suppliers and neither the author nor the publisher makes any guarantee as to the safety of these products or their suitability for solar observing. Try to solicit the advice of an experienced solar observer before buying any piece of equipment.

Suppliers of astronomical equipment exist in vast numbers nowadays. I have therefore restricted this list to those companies selling products either specifically for solar observing or useful for solar work. To find suppliers of telescopes and other general astronomical equipment, peruse the advertisements in the popular astronomy magazines (see Appendix C, "Further Reading").

Suppliers in UK

Beacon Hill Telescopes, 112 Mill Road, Cleethorpes, Lincolnshire DN35 8JD. Suppliers of Baader AstroSolar Safety Film as well as a wide range of telescopes and many useful accessories, such as camera adapters.

Broadhurst Clarkson and Fuller Ltd, Telescope House, 63 Farringdon Road, London EC1M 3JB; www.telescopehouse.co.uk. White-light solar filters plus many telescopes and accessories by leading US manufacturers.

David Hinds Optics, Unit 34, The Silk Mill, Brook Street, Tring, Hertfordshire HP23 5EF; www.dhinds.co.uk. Suppliers of Baader AstroSolar Safety Film white-light solar filters; also of telescopes and accessories by Celestron and other major manufacturers.

Venturescope, The Wren Centre, Westbourne Road, Emsworth, Hampshire PO10 7RN; www.venturescope.co.uk. Suppliers of H-alpha filters and Baader Planetarium coronagraphs, as well as numerous brands of telescopes and accessories.

Suppliers in North America

Astro-Physics Inc., 11250 Forest Hills Road, Rockford, IL 61115, USA; www.astrophysics.com. US agents for equipment by Baader Planetarium. Also manufacturers of high-quality refracting telescopes and mounts.

Celestron International, 2835 Columbia Street, Torrance, CA 90503, USA; www.celestron.com. Solar filters and accessories to suit the wide range of Celestron telescopes. Celestron do not sell directly to the public, but their products can be obtained through Celestron dealers worldwide. Contact Celestron to find your nearest dealer.

Coronado Instruments, Showroom 1684 S. Research Loop, Ste. 524, Tucson, AZ 85710, USA; www.coronadofilters.com. Sub-angstrom H-alpha filters for the amateur solar observer. Filters can be obtained either directly from Coronado or from agents across the world.

Kendrick Astro Instruments, 2920 Dundas Street West, Toronto, ON M6P 1Y8, Canada; www.kendrick-ai.com. Suppliers of Baader AstroSolar white-light filters.

Lumicon, 6242 Preston Avenue, Livermore, CA 94550, USA; www.lumicon.com. Supply their own 1.5-angstrom solar prominence filter and many useful accessories.

Meade Instruments Corporation, 6001 Oak Canyon, Irvine, California 92618, USA; www.meade.com. Manufacturers of a large range of telescopes and accessories. Meade do not sell direct to the public, but their products can be obtained from dealers worldwide. Contact Meade to find your nearest dealer.

Orion Telescopes & Binoculars, PO Box 1815-S, Santa Cruz, CA 95061, USA; www.telescope.com. White-light solar filters to fit a wide range of telescopes; also many useful accessories.

Thousand Oaks Optical, Box 4813, Thousand Oaks, CA 91359, USA; www.thousandoaksoptical.com. Manufacturers of both glass and plastic white-light solar filters for visual observing, solar eclipse viewers, glass filters for solar photography and 1.5-angstrom H-alpha prominence filters.

Appendix B
Solar Observing Organisations

If you are new to solar observing, and particularly if you are starting out in astronomy, the first place you should go to for advice on observing is your nearest local astronomy club or society. There are bound to be some members there with experience of solar observing and most are willing to provide help and encouragement as well as advice on what equipment to buy. If you cannot find a local society in your area, try contacting the following national sources of information.

(In the UK): The **Federation of Astronomical Societies**, c/o 10, Glan-y-Llyn, North Cornelly, Bridgend County Borough, CF33 4EF; www.fedastro.org.uk.

(In the USA): The monthly magazine *Sky and Telescope* (see under "Magazines" in Appendix C) maintains a list of astronomy clubs and also planetariums and museums.

National Organisations in UK

The leading association of amateur astronomers in Great Britain is the **British Astronomical Association**, Burlington House, Piccadilly, London W1J 0DU, UK; www.britastro.org. The BAA publishes the authoritative *Journal of the British Astronomical Association* and has a very active Solar Section, to which several dozen members send monthly observations. You can obtain details on the Solar Section by writing to the BAA at the above address or through the BAA Web site.

If you are new to solar observing you may wish to consider joining the **Society for Popular Astronomy**, 36 Fairway, Keyworth, Nottingham NG12 5DU, UK; www.popastro.com. The SPA is a national society for beginning and intermediate-level amateur astronomers of all ages. Its Solar Section helps beginners master the basic techniques of solar observing. The SPA also publishes a lively magazine, *Popular Astronomy*, four times a year, as well as regular News Circulars which feature observational reports sent in by members.

The Astronomer, 6 Chelmerton Avenue, Great Baddow, Chelmsford, Essex CM2 9RE, UK; www.theastronomer. org/index.html is a monthly magazine dedicated to publishing astronomical observations and discoveries quickly. It has an active Solar Section.

National Organisations in North America

The **American Association of Variable Star Observers**, 25 Birch Street, Cambridge, MA 02138, USA; www.aavso.org, has a Solar Division which focuses on monitoring the level of solar activity using sunspot counts supplied by contributing members.

The **Association of Lunar and Planetary Observers**, PO Box 13456, Springfield, Illinois, 62791-3456, USA; www.lpl.arizona.edu/alpo, also has a Solar Section which is especially notable for recording solar activity pictorially – by drawings, photography and electronic imaging.

Royal Astronomical Society of Canada, National Office, 136 Dupont Street, Toronto, ON M5R 1V2; www.rasc.ca. Publishes the very useful annual *Observer's Handbook*.

Appendix C

Further Reading

Magazines

The major commercial astronomy magazines often contain articles on solar observing and the latest developments in our understanding of the Sun. If you are a serious solar observer (or a serious amateur astronomer of any kind), it is a good idea to get at least one magazine regularly, as they often present information that is more up-to-date than in books.

Astronomy Now (PO Box 175, Tonbridge, Kent, TN10 4ZY; www.astronomynow.co.uk), is the main commercial astronomy magazine in the UK and is available from many British newsagents.

In North America there are two major commercial magazines: **Sky and Telescope** (49 Bay State Road, Cambridge MA 02138-1200, USA; www.SkyandTelescope.com) *Astronomy* (21027 Crossroads Circle, PO Box 1612, Waukesha, WI 53187, USA; www.astronomy.com). Both magazines can often be found in British newsagents as well.

Many national amateur astronomy organisations produce magazines of their own. See Appendix B for more information about these societies.

General Books on the Sun

KELLY BEATTY, J, PETERSEN, C C and CHAIKIN C, *The New Solar System*, 4th Edition (Sky Publishing Corporation and Cambridge University Press, 1999).

LANG, K R, *The Cambridge Encyclopaedia of the Sun* (Cambridge University Press, 2001) is a beautifully-illustrated, up-to-date guide to our nearest star. It describes our current understanding of the Sun and the latest research by professional astronomers to a quite high technical level, although mathematics is presented separately in text boxes and so does not interrupt the flow of the narrative.

LANG, K R, *Sun, Earth and Sky* (Springer-Verlag, 1995). A guide to the Sun and its influence on the Earth.

PHILLIPS, K. J H, *Guide to the Sun* (Cambridge University Press, 1992). An accessible but detailed guide to the Sun and how it works.

TAYLOR, P O and HENDRICKSON, N L, *Beginner's Guide to the Sun* (Kalmbach Books, 1995). An easy to read description of the Sun and solar research, with a short chapter on solar observing for amateurs.

Books on Observing the Sun

BECK, R, HILBRECHT, H, REINSCH, K and VOLKER, P (Eds.), *Solar Astronomy Handbook* (Willmann-Bell, Inc., 1995). A very detailed compendium of amateur methods of observing and imaging the Sun, written by many authors and covering a number of technical topics not described in this book, including observing the Sun at radio wavelengths. A good reference for the more advanced solar observer.

KITCHIN, C R, *Solar Observing Techniques* (Springer-Verlag, 2001). A technical guide to observing the Sun, with detailed sections on image processing and observing solar eclipses.

MOBBERLEY, M P, *Astronomical Equipment for Amateurs* (Springer-Verlag, 1998). Includes advice on equipment for solar observing.

TAYLOR, P O, *Observing the Sun* (Cambridge University Press, 1991). A detailed guide to monitoring solar activity by sunspot counting and electronic methods.

Books on Photography and Digital Imaging

BERRY, R and BURNELL, J, *The Handbook of Astronomical Image Processing* (Willmann-Bell, Inc., 2001). A detailed guide to image processing, with a companion CD-ROM.

CHARLES, J R, *Practical Astrophotography* (Springer-Verlag, 2000). A detailed guide to the techniques of amateur astrophotography.

COVINGTON, M A, *Astrophotography for the Amateur*, 2nd Edition (Cambridge University Press, 1999). Another detailed guide to all aspects of amateur astrophotography.

DRAGESCO, J, *High Resolution Astrophotography* (translated by Richard McKim, Cambridge University Press, 1995). An authoritative and very readable guide to photographing Solar System objects at high resolution, written by one of the world's finest amateur astrophotographers. There is a very strong section on solar photography and the book also contains detailed treatment of the strengths and weaknesses of different cameras, telescopes, films and techniques. Essential reading for the serious solar photographer who has passed the beginner stage.

RATLEDGE, D (Ed.), *The Art and Science of CCD Astronomy* (Springer-Verlag, 1997). A multi-author guide to CCD imaging, which includes a chapter on solar imaging.

Reference Books

The Handbook of the British Astronomical Association is published annually by the British Astronomical Association (Burlington House, Piccadilly, London W1J 0DU). It contains detailed solar data, including P, B_0 and L_0 (essential for working out solar coordinates) tabulated at five-day intervals. See Appendix B for more about the British Astronomical Association.

The Astronomical Almanac is also published annually and is a collaboration between Her Majesty's Stationery Office in the UK and the United States Naval Observatory. It is available in both countries from astronomy book suppliers. The *Almanac* tabulates P, B_0 and L_0 at daily intervals.

Index

Please see page 177 for ordering details.